Visions
of the
Multiverse

By Dr. Steven Manly

New Page BOOKS

A division of The Career Press, Inc.
Pompton Plains, NJ

VISIONS OF THE MULTIVERSE
EDITED BY JODI BRANDON
TYPESET BY DIANA GHAZZAWI
Cover design by Ian Shimkoviak/The Bookdesigners
Printed in the U.S.A.

To order this title, please call toll-free 1-800-CAREER-1 (NJ and Canada: 201-848-0310) to order using VISA or MasterCard, or for further information on books from Career Press.

The Career Press, Inc.
220 West Parkway, Unit 12
Pompton Plains, NJ 07444
www.careerpress.com
www.newpagebooks.com

Library of Congress Cataloging-in-Publication Data

CIP Data Available Upon Request.

To Allison and David.

Acknowledgments

rojects like this don't have a beginning; they evolve from other things. This project, in particular, evolved from a different project that wasn't going anywhere very fast. Every time I commiserated with Gary Heidt about that other project, he would tell me, "Write a book about the multiverse." Finally, I listened. Thanks, Gary. Thanks to Michael Pye and the folks at New Page Books for giving this project a chance. I'm grateful for Kirsten Dalley's enthusiasm and suggestions. Go Columbia! Stephanie Brown Clark, a true friend of the written word, graciously read every chapter and gently provided me with needed feedback and a patient ear. I am grateful for her help and advice. Thanks to Jonathan Sherwood, Susan Gibbons, Connie Jones, Angie Zemboy, and Sylvia Manly for helping me define my voice through feedback on previous projects. Thanks to Alyssa Ney for fun discussions about the strange ideas in this book. Looking further back in time, I'm indebted to Sylvia Manly for her encouragement to write about physics. I ask the forgiveness of my family and my research colleagues for the distraction produced by my living in other universes for the past year. Finally, I would like to thank the many fine physicists and philosophers whose creative ideas, books, and papers allowed me to explore the multiverse.

Contents

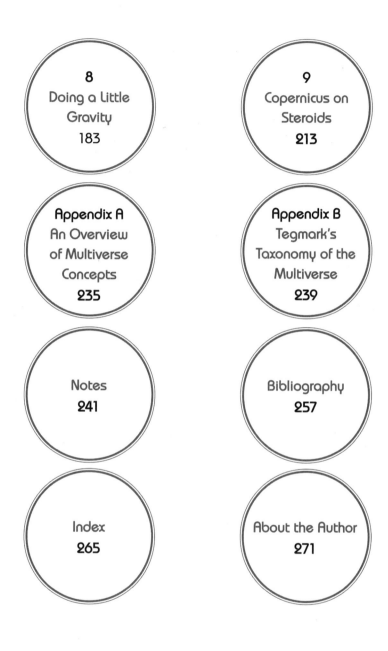

Introduction

Welcome to the multiple universe reality—whatever *that* is. It sounds like something from *Star Trek*. Yet more and more often I see references to multiple universes, or the multiverse, as if it were a respectable idea.[1] The multiverse *is* in *Star Trek*[2] and *The Matrix*, and in literature such as C.S. Lewis's *Chronicles of Narnia* and J.K. Rowling's *Harry Potter* series, as well as countless hardcore science fiction books. Beyond that, though, I'm seeing the multiverse discussed in serious scientific papers by legitimate scientists, some of whom are people I know and respect. Talk about scary!

The whole concept of a multiverse seems rather ludicrous if you think about it for a moment. Isn't the universe all that there is? How then, you might wonder, can there be multiple universes?

Without a doubt, there is a semantics issue surrounding the multiverse. After all, the term *multiple universe* is something of an oxymoron. Forget the idea of multiple universes for a moment and ponder the concept of a single, simple, solitary, stand-alone universe. To me, *the universe* is a term used to denote everything that physically exists. It encompasses the totality of space and time and energy and physical constants and dogs and, yes, even teenagers. It's a rather all-inclusive definition, so it's hard to imagine that there's more to things.

But what does it mean to say something exists physically? The implication is that we can experience it somehow—we can smell it or taste it or ferret out a signal implying its existence using a sensitive scientific experiment. Is it possible to imagine realms that might exist without a physical or causal connection to our known universe? You bet. It *is* possible to imagine such realms and that's where the multiverse comes in.

The idea of multiple universes is nothing new in the non-scientific realm. Consider the Christian concept of Heaven, for example. Admittedly, what is meant by Heaven—and the corresponding price of admission—varies depending on which of the diverse Christian sects we consider. A common theme is that Heaven is a separate plane of existence where a soul enjoys eternal life and pleasure in God's presence and in the presence of other elect souls. Exactly how one becomes one of the elect souls varies from denomination to denomination, but many people believe there's a connection between behavior during one's life and admission to Heaven. Outside of that, there is no connection between Heaven and the reality in which we live our lives on Earth. There is nothing a physicist would consider as a scientific, causal connection between Heaven and our universe. In this sense, Heaven is outside our universe and the two planes of existence—Heaven and the world in which we live—collectively constitute a multiverse.

One problem with the multiverse concept is that there isn't just a *single* such concept. There is a multitude of different types of multiverses. The "Heaven-plus-here" multiverse mentioned is an example of what I call a "faith-based multiverse." There are many other faith-based multiverse concepts, historical and current, such as ancient Egypt's Kingdom of the Dead and Fields of Yalu, and the multiple levels of Jannah in Islam. As I mentioned earlier, there's nothing new in this. What *is* new is that there are a number of serious multiverse concepts coming out of modern physics. Distinct concepts of a multiple universe reality spring from quantum mechanics, cosmology, string theory-based cosmology, and ideas about a mathematics-based reality that shares features with the faith-based concepts.

My aim here is to provide you with an overview of the different multiple universe concepts, along with the background to understand them informally. Along the way, we'll discuss the degree to which these ideas are based on science and the extent to which they are truly revolutionary. Are you ready to journey with me through this strange new reality? If so, read on!

1
Got Copernicus?

Are we in the midst of a fundamental revolution in our worldview? Perhaps. It's happened before. There was a time when the Earth was thought to be at the center of the universe. Then along came a guy named Nicolaus Copernicus. As I heard the story, the revolution began when he was 16 and his frustrated mother took him aside and said, "Nicky, you're driving me nuts! When are you going to realize the universe doesn't revolve around you?"

Ba-da-ba! Feel free to groan. Sorry. There's probably a reason why I do physics instead of stand-up comedy.

What makes this joke funny—to some folks, anyway—is that it plays off the modern folklore surrounding Copernicus. Most of us grew up learning that the ancient Greeks thought of Earth as the stationary center of the universe with all the heavenly bodies embedded in perfect, nested, transparent spheres rotating around the Earth. After modifications by Ptolemy and others to include epicycles in around 100 AD,[1] this cosmology worked fairly well in describing the motion of the sun, moon, planets, and stars. This so-called Ptolemaic view of the universe is said to have made the Christian church happy, because man has a special place in the universe—at the center. In the 1540s Copernicus came along and, apparently having worked through his selfish teen years, determined that a sun-centered or heliocentric universe accounts for the observations with far less complexity than the Earth-centered picture. Soon afterward, Johannes Kepler discovered that a heliocentric universe with the planets moving in slightly elliptical orbits does even better in describing the motion of the heavenly bodies. Eventually, more and more precise measurements and work by Galileo and Newton showed that the Earth-centered universe does not account for the observations nearly as well as the heliocentric model. The Church chose to fight Copernicus and the heliocentric universe because the demotion of mankind from the central place in the universe was theologically unpalatable. Proponents of the heliocentric point of view were persecuted. Galileo was found guilty of heresy and placed under house arrest by the Catholic Church, for example. Eventually, the overwhelming weight of scientific evidence overcame the objections and the heliocentric universe became accepted generally.

This "Copernican revolution" is held up as a great, dramatic victory for the scientific method and lauded as the first in a series of fundamental scientific advances springing from the demotion of the importance of the human perspective and human place in the universe, or what might be called the "human bias." The great late-19th-century and early-20th-century advances in science such as relativity, quantum mechanics, molecular biology, and evolution demonstrated clearly the power of the scientific method and the folly of adhering too much to the human bias. Long live the Copernican revolution!

As with much of popular history, the encapsulated version of the Copernican revolution bears little resemblance to what actually happened at the time.[2] According to many ancient and medieval religious scholars, man's place at the center of things—among the muck and the dirt—was fitting, not because of the privileged position, but rather because man is base and sinful. Relative to the heavens, Earth was a gross and inferior place. Many people at the time would have viewed the movement of Earth away from the center as a promotion for man rather than a demotion. Even now, as a modern kid brought up in a region with Western Christian traditions, I was told that Heaven is up and Hell is down toward the center.

Though the Copernican universe *did* present grave theological difficulties to the religious establishment, with some exceptions it wasn't until a hundred years after the publication of Copernicus's heliocentric model that the Church went to war against heliocentrism and began painting the center of the universe as a privileged place. Shortly after Copernicus published his model, heliocentrism found the majority of its critics among the astronomers. Improved observations and elliptical orbits were needed before the heliocentric model was seen as being demonstrably better than the Ptolemaic model in describing the detailed motions of the heavenly bodies. In addition, the promotion of Earth to a place in the heavens in the midst of the planets circling the sun led to enormous conceptual difficulties. Newton had yet to be born, much less formulate his laws of motion and gravitation. What makes the Earth move? How is it we don't fall off the planet? Before Copernicus, massive things were known to fall toward the center of the universe. Why then, if Earth is not at the center of the universe, do objects fall toward the center of the Earth?

Although things were not nearly as clear-cut as popularly portrayed in the muddled version of the historical record, our encapsulated hindsight is true to the essence of the Copernican revolution. Experimental observation and a desire for simplicity led to the adoption of a model that rejected more than a millennium of human bias and greatly advanced our conceptual understanding of the world in which we live.

The rise of the heliocentric universe was only one example of a major advance in science made possible by pushing beyond our naturally anthropocentric view of the world, which is to say, our "human bias." Many of the fundamental scientific paradigm shifts of the past 150 years were enabled by mind-twisting shifts in thinking. Evidence that tiny particles act like waves and that waves exhibit particle characteristics led to quantum mechanics even though these strange facts go against human intuition. The apparent constancy of the speed of light regardless of the point of view of the observer led to relativity and the realization that time is not absolute—all of which is at odds with human biases about the physical world. Evolution is based on natural change and human evolution is predicated on the observation that humans are animals. These are premises that violate the human need for uniqueness inherent in many religions and philosophical viewpoints. For example, members of some Christian sects object to the teaching of evolution. The methodology of science often requires a *different* perspective—one that steps away from the human bias.

If scientific advances often come at the sacrifice of the human perspective, it is not surprising that science and religion are at odds with one another. The conflict between the two has a long and storied history that continues to this day. Though this strife is a very serious business—the persecution of non-believers or disputes over what should and should not be taught in schools—there is nothing fundamental about the business of science or religion that fuels the conflict. The problems arise when people confuse science and religion or insist that they are absolutely mutually exclusive. Confusion arises easily because both science and religion attempt to provide us with a framework with which to understand and cope with the world in which we live. The methodologies of science and religion are *completely different*, however.

Religion is a matter of faith. Religious people believe in, to a greater or lesser degree, stories passed down to them by members of their community or others that they have sought out. Generally, these stories are much more than entertainment, often containing elements of historical fact and providing guidance on moral and social issues. The methodology of religion does not involve experimental observations of the physical world and, though religions do change and evolve, there is nothing inherent in the methodology of religion that drives this evolution other than the need to stay somewhat relevant in the face of a dynamic and changing society.

Science, on the other hand, bows before experiment. Any hypothesis, no matter how dearly held, is subject to being overthrown by an experiment. Scientific concepts and models evolve as new observations and hypotheses are made. The change is not always quick and not always smooth. Mistakes are made along the way. Still, through a progression of experiments and evolving ideas, concepts and theories that are not consistent with nature are thrown out in favor of those that "fit the data."

Is science completely divorced from faith? Not at all. The very success of science, combined with the appeal of the logic of science, leads some people to something of a scientific faith that is not so very different from religious faith. More on this later.

Is religious faith completely divorced from science? Again, not at all. Religion needs to stay relevant in a society that changes with the development of ideas and technologies driven, in a large part, by science. Religions help society deal with these changes or, in some cases, attempt to isolate society from the changes. Other places where religion intersects with science are instances where particular faith-based views are presented as if they are based on scientific methodology. An example of this is "intelligent design."

Regardless of personal feelings about religion, most of us would admit that science is a very successful methodology by most measures. The conceptual framework developed by science has been used to cure diseases, increase the food supply, provide housing and transportation, manufacture MP3 players, and all sorts of other practical and important things. Science has transformed our lives.

These practical achievements, as well as the more amorphous goal of developing an empirically supported conceptual understanding of the universe, have come through careful, systematic, scientific observations of the world around us. In addition to the scientific methodology used in making observations, scientists have developed instruments capable of probing realms of the universe far outside the typical human experience. Powerful telescopes probe outer space, viewing distant objects in many different frequencies, and microscopes, particle accelerators, and many other types of devices probe the depths of inner space.

It's a strange new world in the realms opened up since the time of Copernicus. Experiments in these new realms have pushed us farther and farther away from human bias and our naturally anthropocentric view of the universe. Today modern physics is carrying this Copernican struggle to the ultimate level, where not only is mankind denied a privileged place in the universe, *our universe* is not even special.[3]

The idea that we live in a multiple universe reality springs from many different sources, which makes for a confusing situation for most of us mere mortals. When faced with a bewildering mess, it is helpful conceptually to form categories out of the confusion. Though there are different ways to do this, one useful way of looking at the different multiverse concepts is to categorize them according to how the universes are separated in the cosmic whole. Within the current multiverse concepts, individual universes are separated by space and/or time, faith, or dimensions.

In Appendix A you will find a categorization, or taxonomy, of multiverse concepts. I've included summary descriptions for each of the multiverses in the taxonomy. It's worth a look at this stage, though you should keep in mind that many of the descriptions should make little sense to you now. All of these concepts will be discussed in more detail along with the necessary background to understand them in the subsequent chapters. They are listed in Appendix A for ease of reference later.

If you count, you'll see that I listed 11 distinct types of multiverses distributed among three different categories or classes in Appendix A. Many of these ideas have versions that differ a bit in detail, and in some instances a multiverse concept might fairly belong to more than one of the categories. Though I tried to be inclusive, I've little doubt that the list is incomplete. This is a fast-moving area in physics today, and theoretical physicists are adept at cooking up new ideas. It is likely that between this writing and the time you read this, there will be additions to the multitude of multiverses.

Some of these multiverse concepts are new. Others are old. Some are inferred from solid scientific evidence and others have no scientific basis at all. Some have a basis in our ideas about quantum mechanics; others are more associated with cosmology, though this distinction is less clear than you might think naively, because quantum mechanics and cosmology are intricately intertwined in modern physics (as we will see in later chapters). The relative positioning of these multiverse concepts along axes defining the degree to which the multiverse is scientific or not, or based in quantum mechanics versus cosmology, are given in Figures 1.1 and 1.2. The positioning shown here is qualitative and based solely on my judgment, though evidence supporting the relative positioning will be provided throughout this book.

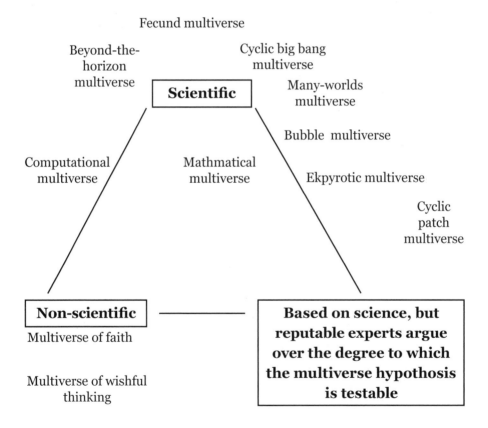

Figure 1.1: Relative degrees to which different multiverse concepts are scientific.

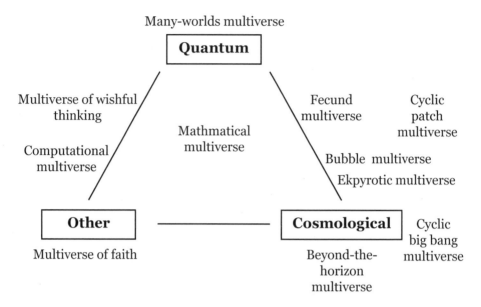

Figure 1.2: Relative degree of quantum versus cosmological character for different multiverse concepts.

The taxonomy presented in Appendix A, where the multiverse concepts are classified according to how the universes are separated in the cosmic whole, is something that I made up as a useful way to introduce order to a messy subject. Think of this as a populist taxonomy. I have included several items in this list that would not be described in serious scientific circles as a multiverse, such as the multiverse of wishful thinking and the multiverse of faith. The multiverse of faith is included because many people believe in other planes of existence, and it is important to draw distinctions between those and the concepts coming out of modern physics. The multiverse of wishful thinking is included because I was surprised when a friend told me that the phenomenal best-seller *The Secret* is based on quantum mechanics. In spite of what the self-help gurus may say, to my knowledge there is no legitimate science in the law of attraction, though I believe firmly that positive and projective thinking helps in the pursuit of life's goals.

A different, and somewhat more formal, taxonomy of the multiverse landscape has been developed and pushed widely by a well-known Swedish-American physicist named Max Tegmark.[4] Armed with a PhD from Berkeley, Tegmark has made a name for himself as a cosmologist. Among other things, he's put forth some interesting ideas concerning the multiverse. Tegmark is currently a professor at an obscure little university in Cambridge, Massachusetts, called MIT.[5]

Tegmark's multiverse classification is a bit stricter in its definition of multiverse than the populist version I floated earlier. According to Tegmark, there are four different levels of multiverses, which correspond roughly to four levels of abstraction. Tegmark's taxonomy is presented in Appendix B. As with Appendix A, these descriptions should be thought of only as a preview of things to come that will be explained more fully in subsequent chapters.

Tegmark's path to the multiverse was far from straight and narrow. In a 2008 interview with Adam Frank, he relates how he started college as an economics major. Only after a friend gave him a book by the famous and fun-loving theoretical physicist Richard Feynman, did Tegmark become curious about physics. Frank quotes Tegmark as saying, "I couldn't understand how this could be the same boring stuff from high school.... If you see some mediocre guy walking down the street arm in arm with Cameron Diaz, you say to yourself, 'I'm missing something here.' So I started reading Feynman's *Lectures on Physics* and I was like, whoa! Why haven't I realized this before?"[6]

With this newfound interest, Tegmark did what any Swedish kid enrolled in college studying economics would do: He enrolled simultaneously in a different college as a physics major! In the interview with Frank, Tegmark admits, "Yeah. You can see I was confused. It got complicated at times. I would have exams in both places on the same day, and I'd have to bike really fast between them."

Tegmark's story sounds a bit like something I read in one of J.K. Rowling's *Harry Potter* books. In fact, much of what you're about to read concerning the multiverse will seem a bit Harry Potterish. You will find it stretching your imagination. A few of the multiverse concepts described in this book are perhaps as fictitious as the *Harry Potter* series, and many of the others are inferred from sound scientific reasoning or educated scientific speculation and, though not proven, and perhaps un-testable, they are worthy of serious thought. One thing is for sure: This is not "the same old boring stuff from high school."

Regardless of how you slice and dice the multitude of multiverse concepts, it is clear that the time has come to take seriously the idea of a multiple universe reality. Is this a fad or a new—and perhaps final—chapter in the Copernican revolution? Read on and see what *you* think.

2
A Brief History of Space-Time

Way back when—more years in the past than I'd care to admit—I was a graduate student in physics at Columbia University in New York City. Predictably, living in Manhattan was an eye-opening experience for a naïve kid from the rural south. That period in my life has provided much cocktail-party fodder through the years. One of my favorite stories from that time began with a loud knock on the door of the apartment I shared with a few other grad students on the edge of the university. When I opened the door, it was a bit of a shock to see a made-for-television cop with credentials held high and a serious look on his face. He asked me if I had noticed a rolled up rug on the street outside the building in the recent past. I answered that I had not. Old furniture and rugs commonly found their way to the street outside our building to await garbage collection and, after living there for a week or so, one ceased to notice it. The next day I picked up the Columbia school newspaper and discovered why New York City's finest was asking about the rug. It seems that a group of freshmen guys had gone out on the town drinking, as students are prone to do. Coming home late at night, they spotted the carpet on the street outside my building. Because the semester had just started and they were still setting up their suite in the dorm, they figured the rug would be a fine addition to their common area. So, the boys lugged the rolled-up carpet to their dorm and unrolled it in their suite, only to discover a murder victim squirreled away in the rug—complete with two bullet holes in the head.

Seriously. It's a true story.

Not to give you the wrong idea. I learned many wonderful things at Columbia, and much of it had to do with physics. Columbia University's physics department has a storied past and many great physicists were there when I was a student, as they are still—though some of the names have changed.

Among the many lessons I learned while at Columbia was that a sense of aesthetics plays a very important role in science. Physicists and mathematicians appreciate elegant mathematics and, oddly enough, it seems that nature is amenable to elegant mathematical descriptions much of the time.

This point was driven home to me one afternoon in Al Mueller's quantum field theory class. Professor Mueller is a highly respected theoretical physicist. In 2003 he won the prestigious Sakurai Prize from the American Physical Society for his work that "...helped to establish QCD as the theory of the strong interactions." This award does not have the cachet of a Nobel Prize in most circles, but, in the world of physics, it's a big deal. Nevertheless, during our field theory class my buddy Lawrence Weincke and I spent much of the time doing what dedicated physics students do during graduate classes: We perfected our impersonations of the professor. When Professor Mueller became very excited about his physics at the board he would go a bit bug-eyed, and his tongue would dart in and out in a lizard-like fashion. We had great fun trying to capture our inner Mueller. On one particular afternoon Professor Mueller was covering the topic of supersymmetry.[1] His back was turned to us as he faced the board writing equations and explaining animatedly some of the fine points of supersymmetry. Suddenly, he spun around doing his most emphatic bug-eye and exclaimed, "This is so beautiful that it... er...well, it just has to be true!" His demeanor—verging on a religious fervor—left no doubt that he believed what he was saying even though there was not one shred of experimental evidence to support the existence of supersymmetry.[2] I left the class with a mixture of amusement and horror, as well as a determination to go learn more about supersymmetry.

Al Mueller hardly stands alone in terms of being a physicist with a deep sense of appreciation of the elegance and beauty of certain ideas and theories. Nor is he the first physicist to hold dear the idea that nature must be described by a beautiful theory. Leonard Susskind, a renowned physicist at Stanford University who is one of the founders of string theory (discussed in Chapter 8), says, "Having spent a good part of a lifetime doing theoretical physics, I am personally convinced that it is the most beautiful and elegant of all the sciences. I'm pretty sure that my physicist friends all think the same."[3]

You might wonder what physicists mean by elegant mathematics or a beautiful theory. That can be a bit fuzzy, because beauty is in the eye of the beholder. Ideas or mathematical structures that exhibit simplicity, symmetry, and efficiency are beautiful. Theories that describe the world around us naturally with the fewest parameters are favored. According to Susskind, "The combination of elegance, uniqueness, and the power to answer all answerable questions is what makes a theory beautiful."[4]

As one of the great theoretical physicists of the day, you might expect Susskind to be progeny of a neurosurgeon and a mathematician or something like that. Instead, he grew up the son of a plumber in a poor family in the South Bronx. In an interview with the *L.A. Times*, he remembered the moment when he informed his father of his career choice: "I was going to engineering school but fell in love with physics. When I told my father I wanted to be a physicist, he said, 'Hell, no, you ain't going to work in a drugstore.' I said, 'No, not a pharmacist.' I said, 'Like Einstein.' He poked me in the chest with a piece of plumbing pipe. 'You ain't going to be no engineer,' he said. 'You're going to be Einstein.'"[5]

Like father, like son? Perhaps. Elegance and beauty in physics are very much the same as elegance and beauty in other arenas. According to Susskind, "[My father] took deep professional pride in finding clever ways to minimize the pipe needed to run a water line from one point to another—without violating the esthetic rules of parallelism, rectangularity, and symmetry…. His pleasure at an ingenious simplification and an elegant geometry was not so different from my own when I find a neat way to write an equation."[6]

Now and then I run into people with the misconception that scientists live in a black and white world. Theories are right or wrong. Concepts are true or false. The cold, calculating Vulcan evaluates the data and spits out facts. The *truth* is that science isn't like that at all. If there is one truth in science, it is that science is a human endeavor. Scientists develop taste in ideas and theories. They like beautiful theories and can have somewhat differing opinions as what constitutes beauty. They can have reasonable but different interpretations of experiments. Mistakes can be made in making measurements. Long-held beliefs often color judgments. Scientific ideas fall in and out of fashion, and careers can be made or broken in the process. Still, don't get the impression that science is a cesspool of disagreements and contentious back-stabbing. It happens, but not that often.

Though it is a human endeavor—with all the confusion, drama, and missteps that are implied—science has a methodology that has enabled incredible advances. The most important thing that sets science apart from all the other endeavors to understand the world around us is that science bows before experiment. Even closely held truths are thrown out if they disagree with experimental observations. Beyond that, the observations and experiments must be reproducible. Additionally, scientists tend to favor the simplest theory that can account for the data. Finally, good scientific hypotheses are required to be falsifiable. If something cannot be tested experimentally, it cannot be put through the full rigors of scientific scrutiny.

Naively, the idea of a multiple universe reality seems to fail as a scientific idea in all respects. By definition, we can't observe a different universe. If we could, it wouldn't be another universe. Also, observations of the universe as a whole are not reproducible in the sense of being able to see a different universe for comparison. We can only see this one universe. As for simplicity, well, the hypothesis of a multiple universe reality seems as if it must be complex as compared to almost any other idea you might put forth. Finally, it's hard to falsify things that, by construction, you aren't supposed to be able to see anyway. So how is it that any concept of a multiple universe reality is taken seriously in scientific circles? That's a good question, and that's more or less the point of this book. Whether or not I convince you in the end that the scientific case for the multiverse is compelling, the truth is that unless you are seeing ghosts or angels, there is nothing in our normal, everyday lives—or even in most of the physics we know— that requires multiple universes. For most of us, our worldview simply does not require more than the one universe in which we live. And that one is weird enough!

Our universe is weird? You bet. Things in our universe are very strange indeed. But to see it we must look at the world around us *very carefully.* That's what physicists do, after all. Because much of this weirdness is critical to scientific multiverse concepts, we're going to take something of a break from the multiverse and examine a few aspects of our single universe in detail.

At first glance, of course, there's nothing weird about our world. Human intuition works quite well in the realm of sizes, times, and speeds in which we live our everyday lives. Even so, the strangeness of reality becomes apparent when we ponder more deeply these most basic of quantities: space, time, and speed.

Though most of us don't lay awake at night fretting about it, we all share the same concept of space and time.[7] From Bill Gates to Sarah Palin to Yao Ming to Joe Smith, we see the world as three infinite and absolute spatial dimensions moving uniformly through absolute time in one direction. If this description doesn't sound all that familiar, allow me a moment to decipher a bit of the physics-speak.

Space is the immutable volume in which we determine *where* things are with respect to one another. Each "dimension" represents a spatial direction where something must be measured to place an event or describe an object. For example, suppose you manage a grocery store and need to tell a helpless customer where to find the corn nuts. To specify the exact position for the customer, you would need to tell him on which aisle he would find the corn nuts (one dimension), how far down the aisle he would find the corn nuts (second dimension), and on which shelf he would find them (third dimension).

Of course, for the truly helpless, even this might not be sufficient. My wife once sent me to the grocery store for panty-liners. Nothing specific. No brand or style preference given, though to know her is to know she would have a strong preference. I found them okay. Without the manager's help, I might add. An entire aisle of panty-liners. There must have been a hundred different brands and styles. Some with wings, others with fresh lemon scent. Alas, it was a traumatic outing. The panty-liner dimension was something my professors in graduate school had neglected in my education.

Time is the fabric in which we determine *when* things are with respect to one another. Without the concept of time, we can't deal with any form of change in the world—and that would be incredibly boring. We know time exists because there is change in our world. According to our intuition, time permeates all space and everyone experiences the passage of time at the same rate. There's no moving forward in time any faster than everything else, and there's no moving backward in time at all.

Saying that space and time are *absolute* is the physics way of saying space and time are the same for everyone, regardless of race, creed, or sexual orientation—and regardless of where they are, how they are moving, and what they are doing at any given moment. An hour passes the same in Shanghai as it does in Rochester. Pilots zipping around in F-15 fighter jets see space and time to be the same as the old codger walking his dog on the street, albeit with better eyesight.

So far, so good. This view of space and time is often called the Newtonian view, because the set of fundamental physical principles formulated using it was first put forth systematically by Isaac Newton in 1687. This won't be the last time we run across Newton's name. He was one of those incredible people who made revolutionary contributions to most everything he touched. Among other things, he invented calculus, dabbled in optics, formulated the basic laws of motion, and put forth a beautiful theory of gravitation that describes everything from falling apples to planetary orbits. To this day, the study of Newton's work constitutes the core of the first term in any introductory university course in physics. Also, like so many of the truly great minds, Newton did more than science. He was knighted by Queen Anne in 1705—not for his scientific accomplishments, but rather for his work late in life as warden of the Royal Mint and Chancellor of the Exchequer (sort of equivalent to the modern U.S. Secretary of the Treasury).

The Newtonian view of space and time—the natural, intuitive sense of space and time that we all possess—is very powerful. Perhaps we are born with this intuition, or perhaps it arises from everyday experience when we are young. Whatever the source, our shared understanding of space and time is needed for each of us to do the simplest of tasks such as walking down a crowded hall, driving a car, or playing basketball. Basic survival requires that we have this intuition. Imagine crossing a busy street or running from a bear if you had no ability to distinguish how fast objects are approaching you. With this handicap, you probably wouldn't last long in the woods *or* the city.

I promised a weird universe, right? Well, in spite of the fact that our basic intuition seems to work, our collective human experience in this great, wide universe is rather limited. Who's to say our intuitive sense of space and time makes sense in all possible situations? Except for a few eccentric physicists, nobody worried much about this until 1905, when Einstein published a paper that probed the nature of space and time in far more detail than ever before. This work revolutionized our view of space and time, and completely flipped out the physicists of the day—and any person learning about it since.

The paper of Einstein's to which I refer is the debut of his famous special theory of relativity. This theory relates observations of events between two observers moving at a constant velocity with respect to each other. Einstein based this work on two fundamental assumptions. The first assumption is that physics is the same for all observers, no matter how they are moving. For example, if an egg drops and breaks, all the observers will see the egg drop and break. There's nothing too strange or contentious in that assumption. The second assumption is that the speed of light is constant for all observers, which is really quite strange when considered in the context of our everyday experience and intuition. Even so, I've found that most people aren't all that bothered by the idea that the speed of light is constant for all observers. It seems that this idea has found its way into our culture.

Why do I say the constancy of the speed of light is odd? Think about it. Imagine going to watch a big parade. Among the marching bands and clowns is a big, long charity float featuring the local star baseball pitcher. On this float is a mockup of a pitcher's mound and home plate. The pitcher is throwing his fastball to the catcher down the length of the float in the direction of travel. Now, suppose you *just happen* to have with you a radar gun that can measure the speed of the baseball with respect to you. True, that's a bit of an odd thing to carry to a parade, but go with it. Let's say that for a moment the parade is stalled and the float stops near you, and you measure the speed of the baseball to be 92 miles per hour—not a bad fastball, and well-known by the local fans to be this pitcher's usual speed. A short while later the float begins moving past you at a speed of 5 miles per hour. Again, you measure the speed of the star pitcher's fastball. This time you will measure the ball to be moving at 97 miles per hour, because the pitcher is moving at 5 miles per hour already when he releases the fastball, and the speeds add together to give the overall speed with respect to you.

There's nothing unusual about additive speeds. If you walk through an airport and encounter a moving sidewalk and step onto it without breaking stride you seem to move faster, and in fact you *are* moving faster. Your speed relative to the building is walking speed *plus* the speed of the moving sidewalk. If you prefer traveling by car, movement on the road is entirely governed by the way in which speeds add. If you are driving at 70 miles per hour and someone in front of you is driving at 55 miles per hour, you'll catch up to them with a speed of 15 miles per hour. The way in which speeds add is an integral part of how motion works in Newtonian physics, and it is something that we all understand intuitively even if we don't spend our days working with the equations.

Okay, back to light. Suppose the baseball pitcher is snapping photos of the catcher rather than throwing baseballs. When the float is stopped, the pitcher takes a photo and you measure the speed of the light emitted by the camera flash. Never mind that your radar gun can do no such thing; this is a thought experiment. We'll pretend the radar gun is some fancy, souped-up version that can measure the speed of light—after all, you ordered it from a late-night infomercial along with a Ginsu knife for three easy payments of $19.95! But wait, there's more! When the pitcher is on the stopped float, at rest, you measure the speed of light to be 670,616,630 miles per hour. This is a *huge* number. It's such a large number that it is difficult to imagine just how fast light moves. To put it in terms that might mean something to most of us, light travels a distance equal to that between New York and San Francisco approximately 64 times in a single second.

As an aside, physicists are a lazy bunch in some ways. As odd as it might seem, they don't like to remember big numbers and lots of equations. So, they make up simple symbols and create mathematical notation that greatly simplifies the many equations that are used to describe our world. The cost of this laziness is that the language and math used in physics is opaque to the non-experts—almost like a secret code that takes years to learn. Once known, the language of physics is incredibly efficient. To be fair, I should not pin this all on laziness. Mathematics is a very rich and powerful language for describing nature, and it can bring to light relationships that cannot be uncovered in other ways. At any rate, I bring this up now because physicists don't like saying "670,616,630 miles per hour" all the time, so we refer to this number as "c."

Now suppose the big baseball float moves along the parade route at 5 miles per hour and the pitcher takes another photo of the catcher. This time, just like with the baseball, you *should* measure the speed of the light to be the inherent speed of light, c, plus the speed with which the float is moving, or c+5 miles per hour. Amazingly, what you measure for the speed of light emitted by the pitcher's camera is c, no matter how fast or in what direction the float moves. In case this doesn't bother you, let me put it in terms of everyday life. This would be as if there is a little old lady driving down the road at 35 miles per hour, and no matter whether you stand still or chase her on a bike or in a car, you will always observe her speed with respect to you to be 35 miles per hour. It's crazy!

You'd think a guy as smart as Einstein wouldn't put such a wacky assumption as the constancy of the speed of light at the core of his theory. Of course, he had his reasons.[8] Perhaps one of them was that two physicists, Albert Michelson[9] and Edward Morley, had observed this constancy of the speed of light in 1887 in a series of very precise experiments. This is an example of the methodology of science at work. In the end, experimental observation trumps expectations, even if it seems a bit wacky.

Our view of space and time was revolutionized by the special theory of relativity. It's easy to see why this is so if we stick with Einstein's two basic assumptions and take another little mind trip to the parade. Imagine that the baseball pitcher on the float becomes bored of tossing baseballs and lies flat on his back on the surface of the float. He points his camera straight up and snaps a photo of the bottom of the banner hanging a few meters above his head that proclaims "Our star pitches for charity!" Because our baseball pitcher is a physics geek at heart, he measures the amount of time it takes for the light to travel from his camera to the banner and back to his eye. From the pitcher's point of view, the light goes straight up and down, as illustrated in Figure 2.1(a).

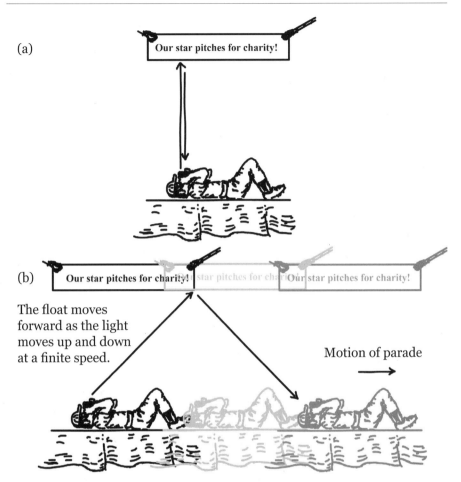

Figure 2.1 a/b: Figure (a) illustrates how the light is perceived to go straight up and down from the point of view of the pitcher, who is riding on the float. In (b), the light from the flash of the camera is perceived to go in a triangular path from the point of view of the spectator standing on the sidewalk as the parade passes. The speed of the parade float is drastically exaggerated in this sketch.

Admittedly, because light moves so fast it isn't practical for the pitcher to time the travel of the light pulse in this situation. That's okay. Because this is just a pretend scheme to illustrate a point, we are allowed to assume the pitcher has Superman-like reflexes and a perfect stopwatch.

Where this thought experiment becomes interesting is when we compare the observation made by the pitcher to what is seen by someone in the crowd off to the side of the parade route. To the parade spectator, the float moves forward as the light pulse travels up toward the banner. Again, because it is a thought experiment, we're allowed to exaggerate the effect of the float moving in order to understand the essence of what is happening. After the light is reflected from the banner and travels back down toward the eyes of the pitcher, the float continues to move forward. So, from the point of view of the spectator, the light pulse seems to follow the path shown in Figure 2.1(b).

Because the height of the banner above the pitcher is the same in both cases, comparing the paths in Figures 2.1(a) and 2.1(b), the spectator sees the light pulse travel along a longer path than that seen by the pitcher. Because the speed of the light (defined as distance/time) is the same in each case and the distance the light travels is farther in the case of the spectator, it must be that the spectator perceives the time for the light to travel along its path to be larger. In other words, time moves at a different rate for the spectator than it does for the pitcher.

What we learn from this thought experiment is that time is relative, not absolute! The special theory of relativity is a framework that relates observations between different points of view (or frames of reference), and, as strange as it seems, this variance in the flow of time depending on relative motion is well established experimentally. Also, the measurements agree very well with what is expected from Einstein's special theory of relativity. Fortunately, the perceived differences seen by observers moving with respect to one another are very small unless the relative motion between the observers is near the speed of light. Time *seems* absolute to us simply because we don't encounter such large relative speeds in everyday life. So, unless the auto companies develop much faster cars, there's no need to adjust your watch according to how fast you drive during your commute.

The relative nature of time leads to all sorts of strangeness. For example, time travel into the future is quite possible, albeit rather impractical. Suppose a starship travels past the Earth at a speed 98 percent that of light. Imagine that astronomers on Earth equipped with a very powerful telescope observe the celebration of the birth of a child on the starship—maybe seeing blue, "It's a boy!" balloons in the window—and mark the date on a calendar. According to relativity, if that space-child grows up and lives what passes for a normal life of 70 years on the starship, the astronomers on Earth will see the wake for that person—black balloons, of course—taking place on the starship 350 Earth years after the birth was observed. Time will pass normally on the starship and on Earth. Nevertheless, when the flow of time between the two reference frames is compared, five years pass on Earth for each year on the starship. If the relative speed between the two points of view were larger, the asymmetry in the flow of time would be even greater.

Time isn't the only seemingly absolute quantity dethroned by relativity. Einstein showed that space is not absolute as well. Measurements of distance depend on one's point of view. For example, if our friends on the starship attempt to measure the length of a football field as they pass the Earth at 98 percent the speed of light, they will perceive its length to be 20 yards (assuming they are traveling in a direction along the length of the field from end zone to end zone). All the markings for the full field will be present, but they will perceive the dimension along their direction of motion to be compressed by a factor of five!

Within the theory of relativity, the set of mathematical formulas that relate what is seen by different observers is called a "Lorentz transformation."[10] The strange effects seen in time and space are included in these equations. What is even more bizarre about these relativistic transformations is that, within them, space and time get mixed up mathematically. This plays havoc with our intuitive view of simultaneity.

To appreciate this, let's imagine there is a big celebration in the United States and massive fireworks displays are ignited near each major city in the country at exactly the same time—say midnight Eastern Standard Time. Correcting for time zone differences means the fireworks are set off in Chicago at 11 p.m. and in San Francisco at 9 p.m. To an observer floating out in space at rest with respect to Earth, it would seem as if sudden blossoms of light spring up simultaneously around all the cities in the U.S. region of the globe. This is all well and good, and probably quite pretty. Now let's ask how this big event would look to our space-faring travelers zipping by Earth at 98 percent the speed of light. In the Lorentz transformation that tells us what the starship crew would see, the time observed in the starship for a particular fireworks display depends not only on the time the display takes place on Earth but the *position* as well. So, the starship crew would not see the fireworks displays as simultaneous, but rather springing up at *different* times depending on the location of the fireworks.

This mixing of space and time between reference frames in relativity shows that space and time are not independent entities. This is why physicists speak about space-time rather than space and time separately. It seems that space and time are intimately entwined in spite of our intuition telling us otherwise.

Now, as important as they are, there's more to life than space and time—at least for those of us who are not string theorists.[11] A multitude of other quantities can be measured, such as force, energy, mass, electric field strength, momentum,[12] and so forth. Because any of these quantities might be measured by observers moving with respect to one another—or, as a physicist would say, observers in different reference frames—it is important that we be able to relate the observations between such observers. The special theory of relativity provides a framework to do that. The strangest aspect of these relativistic transformations is that, as with space and time, the equations mix up pairs of different quantities. When we translate the observations in one reference frame to what an observer in a different reference frame would perceive, we discover that things previously thought to be independent share a deep relationship. One of the most important instances of this kind of thing—and certainly the most famous example—is the way in which mass and energy get mixed up in Lorentz transformations. This is the origin of what is most likely the best known equation of all time, $E=mc^2$, which says that energy and mass, just like space and time, are different faces of the same thing.

Einstein's discovery of mass-energy equivalence appeared in 1905 in one of four papers he published that year in a German scientific journal called *Annalen der Physik*. He was the tender age of 26 at the time. Though any one of these four 1905 papers would have been viewed as an important piece of work, together they catapulted him from relative obscurity to prominence in the field of physics.[13] In fact, the combined import of Einstein's work that year has led many people to refer to 1905 as *annus mirabilis*, a Latin phrase meaning "year of wonders." In one of the four papers, Einstein irrevocably revolutionized our view of space and time[14] with his special theory of relativity. In another, he explained a phenomenon called Brownian motion using the concept of atoms. This work established the existence of atoms in a direct way that had eluded scientists up to that point. In a third 1905 paper, Einstein explained a phenomenon called the photoelectric effect using the hypothesis that light has particulate characteristics. This paper helped launch quantum mechanics. A fourth 1905 paper from Einstein is the one that dealt with mass and energy in special relativity. This work launched the nuclear age and led to other fascinating consequences as we will see throughout this book.

But wait, there's more! You'd think overthrowing humanity's view of space and time, launching quantum physics, demonstrating the existence of atoms, and triggering the nuclear age might have been enough for one person. Apparently not. Einstein published a revolutionary new theory of gravitation in 1915 that he called the general theory of relativity. Since its inception, this theory has provided the framework for physical cosmology, the study of the large-scale structure and evolution of our universe.

The general theory of relativity is similar to the special theory of relativity in that it relates observations between observers in different frames of reference. The "special" in the special theory of relativity refers to the constraint that the observers are moving at a constant speed. In physics-speak, we say that special relativity works only for non-accelerated, or inertial, reference frames. The general theory of relativity basically solves the same issues as the special theory of relativity does, except that it also includes accelerated reference frames. The general theory of relativity is, uh, er, well, more *general* than the special theory of relativity—and aptly named!

Okay, so it's fair to ask, accelerated or non-accelerated frames, what's the big deal? Remember: In 1905 Einstein was on a hot streak. So why did it take him 10 years to make this seemingly tiny improvement in his theory? And what in the world does this have to do with gravity?

As we've already seen, Einstein was a master at noticing the obvious, assuming it as a hypothesis and following the consequences to the bitter end. In the case of general relativity, Einstein noticed that *acceleration is indistinguishable from gravity*. Think about it. It's not as crazy as it sounds. When you stand in a windowless elevator, you feel heavier than normal when the elevator accelerates upward. To your feet, it makes no difference whether you gained a few extra pounds or the elevator floor is accelerating upward. In both cases, the floor presses with more force on the bottoms of your feet. This works the other direction, too. If the elevator accelerates downward, you feel lighter. Momentarily, it feels as if you are standing still and the gravitational pull of the Earth on your body is reduced.

True to his style, Einstein made this obvious similarity between acceleration and gravitation the cornerstone of his general theory of relativity. He hypothesized that the fundamental essence of the gravitational force and acceleration are indistinguishable. Physicists call this hypothesis Einstein's "equivalence principle." Through this hypothesis, the general theory of relativity became much more than a way of relating observations between observers in different reference frames, as it is also a theory of gravitation.

Though the equivalence principle may seem quite strange, there is nothing to keep us (or Einstein) from assuming it to be true in order to explore what follows from that assumption—in essence, playing "pretend" and looking to see what happens. So, let's do just that. Imagine you are standing in a small room with no windows, such as our now-familiar elevator car. What Einstein says is that, short of looking out of the elevator, you can't tell whether you are standing in a motionless elevator on Earth or in a small room on a rocket ship accelerating at a rate of 1g out in space in just such a way to make the floor press against your feet same way that it does on Earth.[15] Think of the rocket ship as a super elevator. As the rocket elevator accelerates upward, you feel pressed toward the floor. We have to play this pretend game with a rocket ship rather than a normal building elevator because we want to consider the case where you are far away from the Earth and any other large source of gravitation. On the following page, Figure 2.2 shows the two indistinguishable cases we want to consider: one on Earth and one in an accelerating rocket. The force of the floor on your feet, which is what you perceive as your weight, is the same in both cases.

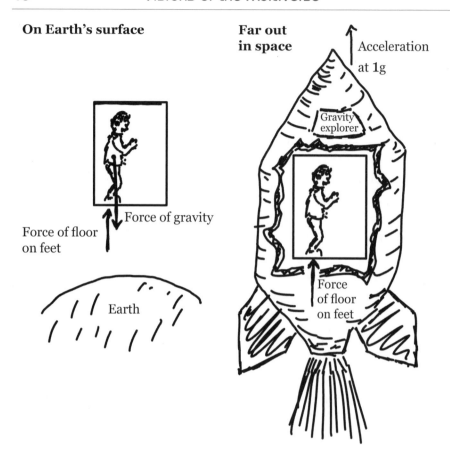

Figure 2.2: According to Einstein's equivalence principle, the force of the floor on your feet on Earth is indistinguishable from the force of the floor on your feet in a system accelerating at 1g far away from Earth and any other planet or star.

Now, following Einstein's lead, let's assume the gravitational force of the Earth on you as you stand in the motionless elevator at the surface of the Earth is indistinguishable from standing in the rocket ship accelerating out in space. To get some idea of the strange implications of the equivalence principle, let us also assume that you have a laser pointer, point it straight at one of the walls of your little room, and turn it on. For simplicity, let's say the light from the laser is emitted as a pencil-thin burst, so that the light is more of a narrow pulse, or bullet, of light rather than the usual continuous beam emitted by typical laser pointers. Consider the path taken by the light from the laser as it heads toward the wall. Whether you are in the elevator on Earth or in the strange little room on the rocket ship, you would expect the light to travel in a straight line from the camera to the wall, right? Sure.

Not so fast. Let's think about the situation we've concocted for a moment. First, let's consider the point of view of someone motionless in space watching the event as the rocket zips by. We'll assume the person looking at the rocket can see through the walls of the rocket and see the pulse of light. After all, it's a game of pretend, right? As the light travels toward the wall of the room, it moves at a finite speed and the rocket (and room) moves upward while the light is in transit. As the light moves a little further toward the wall, the room/rocket moves upward even faster because the rocket is accelerating. This is shown schematically in Figure 2.3(a) on the following page. Now, from your point of view, traveling on the rocket ship watching the same thing, the light seems to curve a little toward the floor as it travels, as shown in Figure 2.3(b), also on the following page. Of course, keep in mind that the extent of the curvature is dramatically enhanced in the figure for your viewing pleasure.

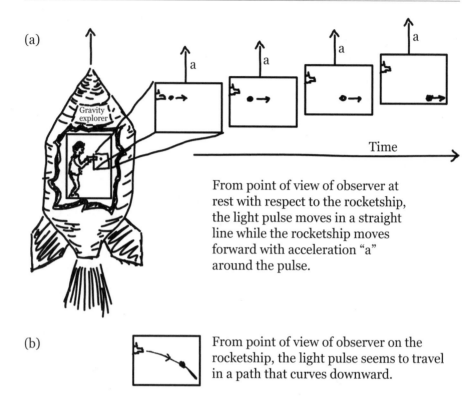

(a)

From point of view of observer at rest with respect to the rocketship, the light pulse moves in a straight line while the rocketship moves forward with acceleration "a" around the pulse.

(b) From point of view of observer on the rocketship, the light pulse seems to travel in a path that curves downward.

Figure 2.3: How the perception of the light pulse depends on the point of view of the observer.

If this little thought experiment made sense to you, then you can accept the fact that the light travels in a curved path as it traverses the little room on the accelerating rocket ship. Now comes the strange part. Einstein's assumption that the accelerating rocket is indistinguishable from the earth's gravitational field comes into play. This means that if you were to shoot a pulse of light across the stationary elevator on earth, the light should also travel in a path that curves downward! What causes the light to curve? Well, it's the accelerating reference frame in the case of the rocket ship and, according to Al, the gravitational force of Earth in the other case. In other words, gravitation can bend light.

Okay, this idea is a little strange. We all grow up thinking of light as traveling in straight lines. In fact, we like to think of a straight line as the shortest path between two points. This certainly works for the world in which we live. It's what we learn in geometry class. But imagine what would happen with this definition for ants in love on the surface of a basketball. As the two ants run toward each other to embrace, the shortest distance between them would be along the curve of the surface of the basketball. Yikes! Welcome to the world of curved space. It turns out that the world which we perceive as filled with straight lines and flat planes is a special case to the mathematicians. They, in fact, have no trouble at all inventing geometries where the planes are curved like the surface of a basketball or a saddle. In such curved geometries, lines follow the planes and are curved.

So, Einstein said that the light crossing the room on Earth *does* travel in a straight line—only the straight line is curved because gravitation bends space. The light still travels the shortest distance between two points, as strange as that may seem. General relativity says that if you shoot light across a room here on Earth, it will move in a path that looks like that sketched in Figure 2.3(b), though the effect isn't noticeable because it is very small.

It's this craziness of interpreting gravitation as curved space that cost Einstein several years as he created his general theory of relativity. He had to develop his theory using the complex mathematics of curved space (known as differential geometry) where things are not quite as straightforward as they are in the Euclidian geometry we all know and love and learn about in school.

In Einstein's theory of general relativity, the essence of gravitation is that it bends space. If that's not weird enough, we already know that space and time get all mixed up when we translate things from one point of view to another, so gravitation can't really *just* bend space by itself; it has to bend time as well. Gravity warps the fabric of space-time and affects the passage of time. Clocks at the surface of the Earth will run at a slightly different rate from clocks on satellites in orbit. Yikes!

Again, as in the case with special relativity, the effects are typically quite small. You won't have to adjust your watch due to the reduced gravitational pull when you drive to the top of a mountain or fly around the country—at least not unless your watch keeps time to an extremely tiny fraction of a second. Nevertheless, the gravitational effects on time and space predicted by Einstein's general theory of relativity are well-measured and quite real. For example, in 1919, Sir Arthur Eddington and two collaborators were the first scientists to measure the bending of light from distant stars around the edge of the sun. This measurement and more modern observations have found this bending of light to be consistent with what is expected from general relativity.[16]

Even though the differences are small typically between general relativistic calculations and those of classical Newtonian gravitation, these differences can be very useful and important in the real world. For example, the Global Positioning System (GPS), which relies on numerous precise clocks placed in space on satellites, requires corrections derived from general relativity in order to attain the precision necessary to be useful. This is a good thing, because the GPS system is quite critical for commercial and military and, increasingly, civilian navigation. I fear that certain unnamed persons in a family with which I am intimately familiar might never make it home from the grocery store without a GPS telling them where to turn. Thanks Al!

There *are* places in this vast universe where the strange effects predicted by general relativity are very large. In fact, we should be very thankful that none of them are in our neighborhood. By far, the crown jewel of weird places is formed when massive stars run out of fuel and cool down. Without the heat pushing outward, gravity pulls the stellar material inward mercilessly, forming an increasingly compact object. As more and more mass is crammed into a smaller and smaller volume, the gravitational force gets stronger and stronger. Eventually, the gravitational force at the edge of the object reaches the point that the space is bent around in a circle as if in the path of an orbit. Light emitted by the object never escapes and, consequently, the object is as black as black can be.[17] Recognizing this in a fit of amazing verbal dexterity, physicists call the beast a black hole.

You might be wondering what in the world all this relativistic stuff has to do with the multiverse. The setting for our universe is made of a fabric of space-time. In many of the multiverse concepts we'll discuss, those other universes are also set in space-time. Relativity is a core part of our understanding of how any scientific concept of the universe or multiverse must work.

Einstein understood this. With his newly developed theory of gravitation and its relation to space, time, and energy, Einstein turned his attention to cosmology, which is the study of the universe and its evolution in time. In 1917, he published a paper entitled "Cosmological Considerations on the General Theory of Relativity" that presented a model of the universe based on the equations of relativity. In this work, he assumed the universe to look the same in all directions and to be homogeneous on large scales. In addition he assumed that the universe should be unchanging in time on large scales, which is to say a "static universe." The reasoning behind this latter requirement was that there was no indication from the astronomers at the time that it might be otherwise, and, philosophically, many people through the years—including Einstein—preferred to think of the universe as having existed forever and likely to go on existing indefinitely into the future.

Modern cosmologists call the type of model that Einstein developed for his 1917 paper a "closed and bounded" universe. In such a theory the matter in the cosmos curves the space in the universe in on itself so that if you were to travel in a straight line you would end up back where you started, as if you were traveling on the surface of a sphere. You could travel forever and never find the edge of the universe, even though if somehow you were able to step outside the universe you would see it as being finite in extent.

To his frustration, the simplest solution Einstein found to the relativistic equations in his model yielded a non-static universe that collapsed under its own gravitational attraction. It's not so hard to imagine, after all. The universe is filled with matter that is attracted to the rest of the matter. How could it not collapse? So, to counterbalance the force of gravity at large distances in the universe, Einstein searched for and found a suitable term that could be added to his equations to provide a repulsive force. This term is known as the *cosmological constant*.

What is the origin of this repulsive force that embodies the cosmological constant? Einstein didn't say. We'll see later that it is still a subject of great interest and debate. I'm getting ahead of myself. For now, keep the cosmological constant in mind. It will come up again and again in our story.

So far we've seen that the world around us seems strange when examined very closely. The oddly counterintuitive theory of relativity that was invented by Einstein to explain this strangeness forms the framework for cosmological theories. As such, it figures prominently in many of the concepts of the multiverse we will see in future chapters.

3

Particles and Fields and Waves, Oh My!

s with the very large and the very fast, human intuition fails us in the world of the very small. And, as strange as it may seem, it is within this realm of the small that we find the seed of a scientifically supported and popularized concept of a multiple universe reality: the many worlds interpretation of quantum mechanics. In fact, quantum processes are of critical importance to all the scientific multiverse concepts listed in Appendix A. Given the central importance of quantum mechanics to our story, I hope you'll forgive me as I drag you through some of the basic ideas underlying the theory.

Quantum mechanics is the theoretical framework within physics where particles and waves collide. To most of us mere mortals, particles and waves are distinct and exclusive concepts. Particles are things like flying bullets or thrown baseballs—easy to localize in space and time. They are here or they are there. Waves are a fuzzier business. Think of waves at the beach. It's easy enough to point out a wave hitting a portion of the beach, but where *exactly* is that wave located? When exactly does it hit the beach? In the world around us, waves and particles have very different characteristics, and that necessarily leads us to a deep reluctance to embrace a theory that sees them as different faces of the same thing. It turns out that nature has no such hang-ups about waves and particles needing to be distinct. So, like it or not, we are stuck with that fact scientifically and must roll with the consequences.

In the late 1800s, physicists had a deep and well-developed understanding of both waves and particles where the two were seen as

47

being quite different, consistent with human intuition. That understanding is still valid today. Particles today, as in the 1800s, move in straight lines unless acted on by a force. Particles collide and scatter. Small, moving masses are easily modeled as particles. Even extended objects such as cars and buildings can be modeled as particles and combinations of particles. For example, you can support an extended object by balancing it at a single point, called the center of mass, as if all of the extended object were nothing more than a single particle concentrated at that point.

Waves, on the other hand, are seen as more of a collective behavior of particles. This is easy to visualize if you imagine being in a huge stadium at a sporting event. The stadium wave seen so often at such events is formed by individuals doing nothing more than standing with raised hands and sitting down. If all the people did that same motion at random times, there would be no wave. Instead, the people in the crowd stand as the crest of the wave passes their section, actually creating the crest of the wave as it passes them. Those people sit and the people next to them stand as the crest of the wave moves on in that direction. Typical waves in nature, such as sound waves, water waves, and waves on vibrating strings, all share the basic characteristic of the stadium wave in that they are made up of particles that vibrate back and forth in a single spot. What makes the wave is the fact that there is a connection between the different particles in terms of the timing and degree to which they vibrate, just as in the case of the stadium wave.

Let's probe waves a bit more and consider what happens when we pluck a taut string, say on a guitar. As we pull on the string we displace a part of the string to one side. Because that part of the string is attached to the bits of the string to either side of where our fingers do the plucking, those bits are displaced as well, just not as much as where we do the plucking. When we let go of the string, the place where we held it snaps back. As this happens, it drags the bits of string to the side with it, and these bits drag the bits next to them, and so forth. The pulse travels down the string in this way. If you were to tie a knot in the string at one point, as the wave passes you would see the knot move up and down in the same way an individual in the stadium wave rises and sits. In both cases the individual bow or person does not move along with the crest of the wave.

Waves are rampant in nature. For waves on a string, the bits of string are displaced sideways relative to, or transverse to, the motion of the wave. In water waves the water moves up and down very similar to what happens in the stadium wave. The displacement in the level of the water takes place transverse to the direction the water wave travels. In sound waves air molecules vibrate back and forth along the direction of the wave travel, forming regions of greater or less pressure. Earthquakes are formed from waves traveling through the Earth's crust. And on and on.

All the waves just described are so-called "mechanical waves." Something—called "the medium" by physicists—has to move for these waves to occur. For example, water is the medium in which a water wave moves. The water moves up and down as the wave passes. For many years, mechanical waves were the only kinds of waves imagined to occur. After all, how can a wave happen if there is nothing mechanical to do the waving? Once again, it seems that nature pays little attention to human intuition. To understand this, we need to consider electric and magnetic forces.

Whenever I think about electricity I always think about getting shocked. In particular, I'm reminded of a kitten I had in graduate school named Shiksa. She was a sweet tortoise shell calico who loved to eat bugs that happened to find their way into my little apartment. That apartment was only slightly larger than the cat's litter box and situated just down the hall from a garbage shoot in an apartment building in New York City. As you might expect, my sweet kitty earned her keep by eating roaches. It kept her entertained and, for all I know, well supplied with essential vitamins. But I digress. I'm reminded of Shiksa when I think of electricity because on cool and dry days whenever I reached out to scratch her head, she would get shocked. Consequently, for several weeks into a cool, dry fall, she habitually cringed away from my approaching hand.

Between cringing kitties, lightning, and spinning compass needles, humans must always have had an awareness of, and respect for, electric and magnetic phenomena. The modern concept of these two forces, now thought to be different faces of a single force called the electromagnetic force, is built on a foundation laid by many people over a long time stretching back to the ancient Greeks. The core of what fills modern textbooks, however, was put forth by James Clerk Maxwell in 1861. Einstein's special theory of relativity cleaned up some important loose ends in Maxwell's theory in 1905, while Paul Dirac, Wolfgang Pauli, Werner Heisenberg, Richard Feynman, Freeman Dyson, Julian Schwinger, Shinichiro Tomonaga, and others succeeded in creating the so-called "quantum" version of the theory in the period roughly between 1920 and 1950.

Electric forces exist between particles that possess electric charge just as there is a gravitational force between particles with mass. The basic way in which the force varies with distance is similar between the two forces as well. Unlike gravitation, however, the electric force can be either attractive or repulsive. Electric charge comes in two varieties and the force is attractive between unlike charges and repulsive between like charges. It also happens that the electric force is about 10,000,000,000,000,000,000,000,000,000,000,000 (or 10^{34}) times stronger than gravitation. Does this surprise you? After all, you spend much of each day fighting gravity and, in the end, in spite of robust underwire bra supports and knee braces, gravity wins if you live long enough. Gravity is the force that dominates all others over long distances and is responsible for the large scale structure of the universe. Still, gravitation is by far the weakest fundamental force known in the universe. The electric force is less noticeable than gravitation in everyday life because the particles that make up matter are typically paired in such a way that the net electric charge is zero. Under certain man-made or natural circumstances the electric charges that make up matter can be separated to some extent and electric forces can be quite noticeable—a fact of which any pet is very aware on cold, dry days.

We know that if two positively charged particles are near each other there is a repulsive force between them. What is it that actually causes that electric force? We'll come back to this question in time, but for now let's imagine that a particle possessing electric charge creates a condition in the space around it, such that if another charged particle is in that space it experiences a force. Physicists call that condition in space an electric field.

The idea of a field is something we'll run into again and again in scientific theories of the multiverse. Think of a field as a mapping of some quantity or characteristic to every point in space. For example, the temperature map shown during the weather report is a representation of a temperature field. You can imagine that every point in space is associated with a number that represents the temperature at that point in space. The map shown during the weather report shows selected temperatures at different locations in order to give viewers an idea of how the temperature varies around the local region at a specific time. The collective set of all the temperature readings and the manner in which they are distributed in space is the temperature field.

In the temperature map example, a single number is associated with each point in space. Fields like this are known as "scalar fields." It turns out that fields can be far more complex beasts than that. Imagine if, instead of making a map of temperature, we chose to make a map of wind speed and direction at every point in space. To do this, at each point we could associate a little arrow that points in the direction of the wind at that point in space and at that time. The length of the arrow would depend on the magnitude of the wind speed at that point—long for a powerful flow of air and short for a small breeze. Physicists can model mathematically little arrows like this, also known as vectors, with three numbers. A field such as this wind field, which consists of a vector associated with every point in space, is known as a "vector field."

The electric field mentioned previously is a vector field. You can visualize it consisting of little arrows at each point of space. Each arrow points in the direction of the force a positive charged particle would feel at that point. The length of each arrow is proportional to the strength of the force felt by the particle at that particular point. Figure 3.1 shows a representation of the electric field surrounding a positive charge. Of course, this sketch is only a two-dimensional representation. The real electric field would be present in three dimensions and have a spherical symmetry. Imagine a beach ball with arrows pointing outward from its surface.

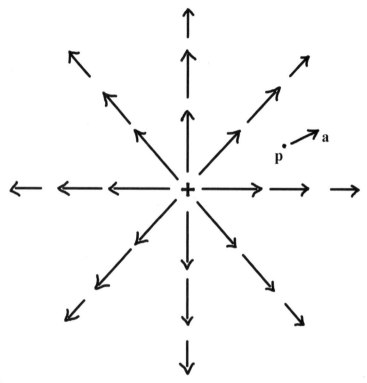

Figure 3.1: A sketch of the electrical field surrounding a positive electric charge in two dimensions. The arrows represent the direction and magnitude of the force that would be felt by a positive "test" charge at each point in space. For example, if a positive test charge were placed at point "p," it would be pushed in the direction of arrow "a" with a force in proportion to the length of that arrow. These are representative arrows. The electric field exists at all points in the space of the paper in this example.

In a fashion analogous to the electric force, physicists imagine that a little bar magnet creates a magnetic field in the space around it and that it is the action of the magnetic field on a compass needle nearby that causes the needle to move. As with the electric field, the magnetic field is a vector field, meaning it has both a direction and a magnitude at each point in space, and is also best visualized as a field of little arrows.

In the typical human experience day-to-day, we see electric and magnetic forces as being distinct. Static cling, lightning, and shocks on little kitty noses are electric phenomena whereas refrigerator magnets are attached to the metal of the refrigerator door by a very different force. In truth, a closer look shows that electric and magnetic forces are intimately connected. Moving electric charges in a wire create a magnetic field that is quite effective at spinning compass needles. Similarly, swinging a magnet near a wire will cause a force on the charges in the wire that is indistinguishable from an electric field.

Back in 1861, the great Scottish physicist James Clerk Maxwell pulled together the various bits and pieces of what was known about electricity and magnetism under the umbrella of a single mathematical theory and made an amazing discovery—one of the most significant scientific discoveries of all time. Maxwell discovered that if a changing electric field exists in some region of space—say, around a wiggling electric charge—it will create a changing magnetic field in the space around it. That changing magnetic field will, in turn, create a changing electric field in space that will create a changing magnetic field, and so forth. This strange series of oscillating fields inducing fields is a wave phenomenon that propagates outward at the speed of light and, according to Maxwell, *is* light. In his words, "...we can scarcely avoid the conclusion that light consists in the transverse undulations of the same medium which is the cause of electric and magnetic phenomena."[1]

Maxwell uncovered the essence of light as a wave of oscillating electric and magnetic fields. Note that he says "undulations of the same medium." It wasn't until Einstein's development of the special theory of relativity in 1905 that physicists realized a medium is not required in order for light to propagate. Light is a very strange wave in that way. It is not formed from a waving physical medium, but rather from the waving of the electric and magnetic fields, which can exist in a vacuum. Still there is little doubt that light is a wave phenomenon. For starters, Maxwell's mathematical treatment shows that the electric and magnetic oscillations are described by equations that are very similar to the equations that describe other types of waves. More than that, though, Maxwell's equations can be used to derive other properties of light that are shared by all the other types of waves that are known. These properties are the things that make waves, er, well, waves.

Maxwell was a great physicist, generally recognized to be in the same league as Newton and Einstein; and like them, Maxwell achieved extraordinary things in multiple arenas.[2] In physics, in addition to uncovering the secret of light and the electromagnetic spectrum, Maxwell revolutionized statistical mechanics and developed techniques for determining the bulk properties of gases from first principles. In physiology, he helped prove that the human eye has three types of receptors sensitive to red, green, and blue light, respectively. In astronomy, he proved that Saturn's rings are neither solid nor liquid, but rather made of many particles. He also suggested a technique for, and directed the production of, the world's first permanent color photograph. All this and he died when he was only 48.

Amusingly, Maxwell, one of the great theoretical physicists of all time, was nicknamed "Dafty" by his schoolmates.[3] His speech came in gushes interspersed with uncomfortably long hesitations, and his classmates thought him a genuine rustic halfwit. Of course, Maxwell was far from being a halfwit. He ended up at Cambridge, where he quickly became well known. According to Goldman, referring to the Maxwell's years at Cambridge:

nother piece of research which entered college legend was Maxwell's experiments to test the ability of cats to land on their feet even though dropped on their back. As Maxwell pointed out, to his critics who branded him as a vivisectionist, the point is to see how quickly cats can do it, which therefore involves dropping them from as low a height as possible. He dropped cats on to his bed from the great height of two inches—and they landed on their feet![4]

Cats notwithstanding, Maxwell's greatest legacy is the electromagnetic wave. What kinds of properties define a wave? All waves, including light, exhibit the phenomena of interference, diffraction, and refraction. Interference is the way in which separate waves add together. For example, when the crests of two waves meet, they form a crest that is the height of the two individual crests added together. On the other hand, when a crest meets a trough, they cancel out momentarily. Diffraction is the way in which waves spread out when they go through small openings. Refraction is the way in which waves bend when they pass through different media, such as the bending of light at the interface of air and the glass of a lens.

Light exhibits each of these properties, and they are all contained efficiently within the mathematics of Maxwell's equations. Most physicists I know consider Maxwell's theory to be a thing of great beauty. Sadly, it's a beauty that takes a certain level of training in the language and concepts of mathematics to appreciate fully. It's a great moment for a physics professor when you see the comprehension dawn on the faces of your students as you cover Maxwell's equations and the nature of light. It's as if they are glimpsing a rose for the first time.

I've done our friend Maxwell a disservice. He discovered much more than the secret of light, assuming that the word *light* refers to visible light, meaning that which humans can see. Maxwell found the key to a broad spectrum of electromagnetic waves of which visible light is just a tiny part. The essence of visible light is no different from the essence of radio waves, microwaves, x-rays, and so forth. What distinguishes these different beasts is how many times the electric and magnetic fields wave back and forth in a second. Physicists refer to this as the frequency of the wave. Closely associated with the frequency is the wavelength of the wave, which is the distance between the successive crests of the wave. Figure 3.2 is a sketch of the spectrum of electromagnetic waves showing the range of wavelengths associated with each part.

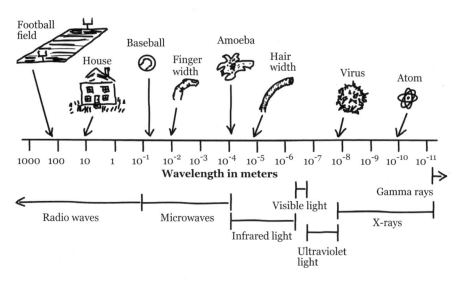

Figure 3.2: The rich spectrum of electromagnetic waves shown as a function of wavelength in meters.

Visible light, gamma rays, x-rays, radio waves, ultraviolet radiation, infrared radiation, and microwaves are all electromagnetic waves. They differ only in frequency. Each of these categories is made up of a range of frequencies of electromagnetic waves. Visible light, which is very dear to us, exists in a range of colors, or frequencies, from red to violet. If your eye could perceive all frequencies of electromagnetic waves, you would see each of these various bands of frequencies simply as different colors. Of course, because humans can see only a limited range of frequencies, much of what happens in the world is invisible to us.

One of the most exciting aspects about the fact that electromagnetic radiation exists in a huge continuum of wavelengths and frequencies is that the waves of different frequencies interact with matter in very different ways. For example, visible light reflects off the surface of your body, whereas x-rays are much more penetrating and can be used to photograph what is *inside* your body. Similarly, visible light cannot be used for radios or cell phones for the simple reason that it will not pass through walls. But, radio waves and microwaves will pass through most walls; these are the frequency ranges used for most wireless communication. Because of these variations in the interaction with matter, the different frequencies of electromagnetic waves are used for different things technologically: gamma rays for radiation treatment, x-rays for imaging inside objects, radio waves for communication, microwaves for radar and cooking, and the list goes on and on. The appreciation of this richness was a great advance in human knowledge that has driven the development of much of the technology and scientific instrumentation surrounding us today.

There is a point to this digression. We will be using the properties of light to probe the nature of quantum mechanics, which is a critical component of scientific models of the multiverse. In addition, light in all its varied frequencies is the messenger from distant parts of the cosmos that we use to infer the large-scale structure and evolution of our universe and surmise the existence of a multiverse. Bear with me.

So, it seems that light is a wave and that is that. Or is it? In the late 1800s, equipped with this amazingly successful wave theory of light, physicists attempted to understand interactions of light with matter. They studied seemingly simple things like the spectrum of frequencies of light emitted by glowing objects and the frequencies of light absorbed during the passage of light through different types of gas. For many years they met with frustration as they were unable to understand their observations within the context of the theories of light and atoms at the time.

Then, in 1900, a German named Max Planck succeeded in describing theoretically the light emitted by glowing objects. The word *glowing* here means only the light *emitted* by the object, as opposed to light *reflected* by the object. As an example of this, think of the glow of a hot oven burner. Things at lower temperatures glow, too, just not as much in frequencies that humans can see. Physicists call this glow—the spectrum of light emitted by an object—*blackbody* radiation. In describing the glow of object, Max Planck started from basic physical principles and derived a mathematical formula giving the brightness of objects in all the different frequencies of electromagnetic radiation—the so-called blackbody spectrum. In order to do this, however, he had to make the assumption that light exists in little particle-like packets, which he called "quanta." The singular of quanta is *quantum*, which is Latin for "how much" or "as much as." For a given frequency, or color, of light, Planck assumed there was a minimum energy that could be emitted or absorbed by an object, that being the energy of the quantum of light. In 1926, well after Planck's work on blackbody radiation, a chemist at UC, Berkeley named Gilbert Lewis coined the word *photon* to refer to Planck's particles of light and the word stuck.[5]

To get a handle on Planck's idea, imagine two siblings fighting over a chocolate bar on the way to soccer practice. Mom, the source of rides and wonderful food, admonishes the owner of the chocolate bar to share. Being no fool, that chocolate-hoarding sibling breaks off a very small bit of chocolate and hands it to her brother. In principle, she could break off an arbitrarily small bit of chocolate and satisfy Mom's request. This situation is sort of like the way people thought about light and matter before Plank. Physicists assumed the energy of light could be absorbed in arbitrarily small amounts. What Plank said was to imagine that the chocolate isn't like a bar of chocolate, but more like M&Ms. Now the stingy sister must share a minimal amount of chocolate in order to satisfy Mom's demand to share. The chocolate comes in quanta of single M&Ms. This is analogous to Plank's idea that light energy comes in little packets that can't be divided.

Planck went further. His model required the photon energy to be proportional to the frequency of the light. In other words, blue light is made of packets with more energy than red light. The chocolate analogy sort of breaks down here. M&Ms of different colors are the same size. To make the M&Ms more like what Planck says for light, we would have to make the blue M&Ms bigger than the red M&Ms.

After Planck released his result in 1900, it was agreed generally that he had succeeded in describing the blackbody spectrum very well. Planck's underlying assumption that light was a particle, however, was viewed as absurd. It was not taken seriously by most physicists of the day. For how could something be a discrete particle if it was known to be a wave? Try to imagine water waves at the beach in terms of small packets. Where are those packets? You can't see them. Water waves look continuous. How can something be broken into packets and yet seem smooth and continuous? The hugely successful wave view of light clearly contradicted the interpretation of light as a particle. Thus, it seemed obvious that Planck got the right answer for the wrong reasons.

This assessment of Planck's work changed radically in 1905 when none other than Albert Einstein used the same idea as that of Planck—that light is absorbed in little packets of energy—to explain a different phenomenon. Einstein used the quantum hypothesis to explain the so-called photoelectric effect, where light incident on a metal surface kicks out electrons from the surface. Like the blackbody spectrum, the photoelectric effect could not be understood in terms of the classical electromagnetic wave theory as put forth by Maxwell, but it was explained quite nicely by the quantum picture of light as put forth by Einstein.

Amusingly, Planck, who won the 1918 Nobel Prize in Physics for this work, thought of pursuing a career in music as he entered the University of Munich in 1874. However, upon asking a musician about the possibility of such a career he was told that if he had to ask the question he would be better off doing something else. Shortly thereafter, he approached a Munich physics professor and was told that a career in physics had a bleak outlook because physics was essentially a complete science with few further developments expected.[6] Fortunately, Planck didn't listen to the latter advice.

The fact that Planck and Einstein were able to explain two distinct phenomena assuming light has a quantum nature leads us to infer that light is a particle. Yet we have already established that light is a wave. Thus it appears that, in spite of the seeming contradiction, light has both wave and particle characteristics. Welcome to the wacky world of quantum physics.

In 1905, quantum physics was similar to an infant: born but not mature. It took time for physicists to accept the child, and it also took time for the child to grow up—decades, in fact. Between 1905 and 1950, the conceptual understanding and mathematical formalism on which modern quantum physics is based—quantum mechanics—was invented. Even with the distinction between continuous waves and discrete particles blurred by the success of the quantum hypothesis for light, it took another 20 years before the next great conceptual step forward was taken in quantum mechanics.

In 1924, in his doctoral thesis, a Frenchman named Louis de Broglie proposed that if a light wave can behave like a particle, then perhaps particles like electrons can possess wave characteristics. By that he meant that particles might have a wavelength and exhibit refraction and interference, and so forth. According to de Broglie's hypothesis, all matter has wavelike characteristics, though typical things in the human experience, such as baseballs, cars, and Twinkies, have wavelengths so short that the wavelike character is not noticeable. For tiny particles like electrons, however, the wavelike aspect is important.

It's fortunate for all of us that history doesn't demand that we refer to our friend Louis by his full name. He came from a family of French nobility that saddled him with the name of Louis Victor Pierre Raymond duc de Broglie. Luckily, he managed to survive the name and a World War I French army stint to win the 1929 Nobel Prize in physics for his proposal that particles have wave characteristics.[7]

De Broglie's Particles-R-Waves hypothesis was a very strange thing to suggest. By 1924, physicists were accustomed to the fact that light seemed to behave like both a particle and a wave. But, you know how it is with the problem child: Nobody's surprised when she shows up with the pierced cheek to go with the pink hair and shaved eyebrows. Light is a wave that can travel through a vacuum unlike all other waves. It's also a wave that can behave like a particle. So light is strange. But to suggest that every other member of the family should go out and get a pierced cheek—well, that's a bit over the top.

Nevertheless, de Broglie's suggestion turned out to be revolutionary. In 1927, American physicists Clinton Davisson and Lester Germer, working at Bell Labs, demonstrated that the scattering of a beam of electrons off a bit of nickel metal behaved as if the electrons were waves with a very small wavelength. Independently, a British physicist named George Thomson similarly discovered the wavelike nature of electrons in his lab at Aberdeen University in Scotland. Thomson and Davisson were awarded the 1937 Nobel Prize in Physics for this work. (In case you were wondering, George Thomson was the son of British physics Nobel Laureate J.J. Thomson. Imagine the conversations around the dinner table in *that* house!)

Is an electron a wave or a particle? And light—is it a wave or a particle? In both cases, the answer depends on the situation. Both material particles and light exhibit wave and particle characteristics. Light acts like a wave as it passes through a lens, but it acts like a particle when ejecting electrons from a metal surface. Strange, but true. The discomfort most of us feel when thinking about this is possibly due to the fact that our imagination and comfort zone are limited by our life's experience in dealing with large objects where the wave and particle world seem well separated. In the world of the very small, the division between waves and particles is less distinct, and things seem very strange to us.

The most critical parts of the mathematical description of matter waves, which we know as quantum mechanics, were conceived in the period just after de Broglie's hypothesis was published. In 1926, an Austrian named Erwin Schrödinger developed a wave equation that could be used to describe matter waves like those de Broglie proposed for electrons.[8] Predictably, this equation is known as Schrödinger's equation. In his 1926 paper, Schrödinger treated the electron not as a particle at all, but rather as a concentrated, or localized, wave, somewhat similar to what you might see if you were holding the end of a jump rope and jerked your hand up and down once. A single wave pulse travels down the rope away from you. To Schrödinger, this single wave pulse was similar to the essence of a particle. It was a picture that deserved serious attention because Schrödinger's solution for the electron in an atom met with striking success in explaining the particular frequencies of light emitted and absorbed by atoms.

In 1925, prior to Schrödinger's work, Werner Heisenberg[9] and Max Born, along with one of Born's students named Pascual Jordan, published a paper that also gave a solution for the electron in an atom based on quantum mechanics.[10] The mathematical techniques used by this trio, however, were found to be unwieldy by many of the physicists at the time. The math was not cast in the wave equation language with which most physicists were so familiar. Of Heisenberg's work, Schrödinger said, "I, of course, knew of his theory but was scared away, if not repulsed, by [the mathematics] which seemed very difficult to me."[11]

In short order, it was realized that the two approaches to quantum mechanics were equivalent. Heisenberg and his colleagues and Schrödinger independently found two ways of looking at the same thing. Modern quantum mechanics was invented twice.

Regardless of how it came about, quantum mechanics á la Heisenberg or Schrödinger is much more powerful than a new model of the atom. Quantum mechanics is quite general. It represents a revolution in how we look at the universe. It can be used, in principle, to solve any sort of physics problem where forces and matter are involved, including many issues in cosmology and, of course, theories of the multiverse. Keep this in mind as we next probe the strangest and most empowering aspect of quantum mechanics: the inherent uncertainty in the theory.

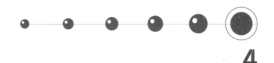

4
God's Gambling Problem

Young students of physics today generally learn quantum mechanics using Schrödinger's approach. The idea that waves are the essential reality of matter—an idea that Schrödinger held dear initially—did not last long, however, at least not with the conviction once voiced by Schrödinger. The central role played by the wave was replaced by what is known as the probabilistic interpretation of quantum mechanics. As we'll see, the revolutionary aspect of quantum mechanics lies more in what it *can't* tell us than in what it can tell us.

For everyday things such as balls and airplanes, we can calculate the future if we know forces and the current state of motion. For example, if you throw a ball and you tell me the speed with which it leaves your hand and the angle at which you have thrown it, I can calculate where it will land using normal Newtonian physics. If you give me very precise measurements of the speed and the angle, and I take into account things such as air resistance and wind, I can calculate the landing point as precisely as you would like. This is how naval ships manage to fire shells and hit targets 20 miles distant.

But, the state of affairs is *very* different if you're dealing with small particles like electrons or photons. "Classical" physics—which is to say Maxwell's equations combined with Newtonian physics—doesn't work for calculating things that involve tiny particles like electrons and photons. That's what led to the development of quantum mechanics in the first place. For small things, we must use quantum mechanics in the same way that we must use relativity to study something that moves very fast. What is really strange, though, is that quantum mechanics doesn't give us the clear answers that we are familiar with from classical physics. Rather than allowing us to calculate the future, quantum mechanics only tells us the *probability* that something will happen. Worse than that, quantum mechanics tells us that it is *impossible* to fully specify the current situation, or in Heisenberg's words, "As a matter of principle we cannot know all determining elements of the present."[1]

To get a better idea about how quantum mechanics and classical physics differ, let's imagine that you adopt a policy of throwing a big party every Thursday. You see one of your buddies and invite him to next Thursday's party. Your friend tells you that they have been invited to several other parties that night. If your friend operates in a "classical physics" sort of way, he would tell you something like "Yes, I will come to your party." Or he might say, "I'll go to another party earlier and show up at your place to end the evening." The point is that your "classical" friend would give you a definite answer. If your friend operates in a "quantum mechanical" mode, you'd hear something like "Well, there's about a 20 percent chance that I'll come to your party." Stranger still, your friend would be *unable* to give you a more definite answer than that. He would only be able to give you a probability that he might attend your party, though that probability might be very well known. So, if your friend was extended an invitation to all of your parties and he said he had a 20 percent chance of making any given party, after a hundred weeks of parties your friend would have attended about 20 of them. As odd as it sounds, there's no way that either you or your friend could know in advance which specific parties he would attend. You would know only that he would be present about 20 percent of the time. Bizarre, huh?!

Quantum mechanics asserts that everything exists in discrete *states* of being. For example, an electron in an atom exists in a particular energy state. All objects, no matter how big or small, exist in discrete states. For big objects, however, such as dogs and golf balls, those discrete states are so close together in energy and distance that you can't tell the states apart. So, for big objects, quantum mechanics looks just like classical physics. Happily, because we live in the world of large objects rather than the realm of atoms, we don't have to concern ourselves with quantum mechanical effects when we go to the grocery store or play football.

For tiny things like electrons, however, quantum mechanics is important. Suppose we have an electron in a certain state of being. It exists in a particular place with a particular energy, and it is influenced by a particular force. We call that state of being the "quantum mechanical state" of the electron. In the Newtonian world with which we are so familiar, if we know the location of something and the forces acting on it, we can calculate where it will be in the future. Quantum mechanics cannot do this. Instead, quantum mechanics tells us the *probability* that the electron will be found in a given state in the future. At some later time, that electron might be found in any of a large number of quantum mechanical states. In principle, all of those states are allowed in that electron's future. Quantum mechanics doesn't tell us in what state we will find the electron when we search for it later. Rather, quantum mechanics provides us with the *likelihood* of finding the electron in any given state.

All this talk of probabilities may sound strange at first, but it's not all that unusual even in daily life. Knowing the probability of something happening is still useful information even if you have no certainty about the outcome. We play the odds all the time in life. For example, most of us are comfortable flying in airplanes only because there is a very small probability the plane will crash.

If you find the fundamental indeterminism of quantum mechanics bothersome, you're in good company. Albert Einstein struggled with this lack of determinism in quantum mechanics. The most famous example of his statements concerning this issue came in a letter to Max Born dated December 12, 1926:

Quantum mechanics is certainly imposing. But an inner voice tells me that it is not yet the real thing. The theory says a lot, but does not really bring us any closer to the secret of the Old One. I, at any rate, am convinced that He does not throw dice.

This last bit of Einstein's letter has been paraphrased and popularized as "God does not play dice with the universe."[2]

In case the loss of our crystal ball isn't enough, the story gets even stranger. Not only does quantum mechanics limit our ability to foretell the future, Werner Heisenberg discovered in 1927 that our knowledge of the present is also fundamentally limited according to the precepts of quantum mechanics. It seems we cannot know to infinitely good precision everything about an object, no matter how carefully we measure it. This is known as the Heisenberg uncertainty principle.

The practical meaning of the uncertainty principle depends on what we are attempting to measure. For example, suppose we're trying to determine simultaneously the position and momentum of a tiny particle. Heisenberg tells us we can't measure the position and the momentum infinitely well. If we measure the position extremely well, the momentum cannot be determined very well and vice versa. This sounds crazy because in everyday experience such a limitation doesn't seem to exist. Of course, our experience is based on measuring very *large* things. Let's think for a moment about a large object, such as a car, zipping down the road in front of us. How do we know its position? We can tell where the car is located because light bounces off the car into our eyes. The car is large, so the process of light bouncing off the car makes no difference in its motion. However, if instead of a car, we consider a small particle like an electron, light bouncing off it will cause it to change direction and/or speed. We can determine where it is at a given moment, but the process of doing so changes its speed and/or direction of motion. Thus, we can't determine very well the position and the momentum of the electron at the same time.

It's difficult to over-emphasize the degree to which the uncertainty inherent in quantum mechanics plays a critical role in modern physics. As we will see throughout much of this book, many of the huge scientific advances of the last century came about because of the liberating nature of these uncertainty relations. These advances also provide the foundation for the scientific concepts of the multiverse that we'll discuss. This prominence has given Heisenberg's uncertainty principle a place in physics culture. Tacked to the wall of the control room for one of the big physics experiments on which I work is a scrap of paper where a student wrote his favorite joke of the day:

> Heisenberg gets stopped in his car by a police officer who says, "Sir, you were speeding. Do you know how fast you were going?" In response, Heisenberg replies, "No, Officer, I don't. But I *do* know *exactly* where I am."

Quantum mechanics was conceived out of the necessity to take into account the wave character of matter. The mathematics centers around the evolution of the so-called wave function of the system. For waves in classical physics (such as waves on strings or in water) the solution to the wave equation is a wave function that describes literally the physical motion of the medium. For example, it describes the position of a bit of the string as a function of time as it vibrates. In quantum mechanics, it isn't possible to interpret the wave function in this way. It isn't clear what it is that's actually doing the waving. In addition, the wave function in quantum mechanics is complex. By "complex" here I don't mean complicated, but rather that it involves real and imaginary numbers. Don't sweat it if the phrase *imaginary number* brings up only a vague memory of something heard in school years ago. The point is that a quantum mechanical wave function cannot be readily interpreted as something material, as in the physical waving of some medium.

Physicists today still argue about the actual physical significance of the wave function. That said, there is little argument about the fact that the strength, or intensity, of the wave function corresponds to probability. For example, the probability of finding an electron at a point in space is greater if the magnitude of the wave function at that point is greater.

The probabilistic interpretation of quantum mechanics was first put forth by the great German physicist Max Born in 1926. His insight into how to think about the wave function along with his other contributions to quantum mechanics earned Max Born the Nobel Prize in Physics in 1954.

Fortunately, the probabilistic interpretation of quantum mechanics is only *part* of Max Born's legacy. He was also the grandfather of the well-known Australian singer Olivia Newton-John. In the lobby of the physics department where I work at the University of Rochester, there are a couple of walls covered with plaques and framed awards received by members of the department. In the midst of these is an autographed publicity shot of the lovely Ms. Newton-John. In the 1950s, Max Born and my esteemed colleague at Rochester, Emil Wolf coauthored a classic textbook on optics. Somewhere along the way, several editions of the textbook later, Emil charmed Olivia into giving him the most striking item on our departmental brag wall. In spite of some fairly nice awards on the wall, it seems to me that Ms. Newton-John's photo garners the most attention from visitors. Go figure.

In spite of the enormous success of quantum mechanics as a scientific framework describing a vast array of complex phenomena, there has been an ongoing discussion among physicists and philosophers about how to interpret aspects of the theory since its inception. In particular, quantum mechanics operates in an indeterministic world of probabilities whereas, in the real world, observations are made with great certainty. It's unclear exactly how this transition from unobserved potential to observed certainty happens and what it means for the fundamental nature of our universe.

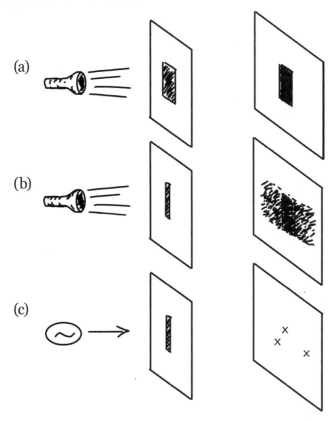

Figure 4.1: In (a), light shines on a large opening in an opaque wall. The image on the screen behind the opening is sharply defined and looks like the opening. In (b), the same light shines on an opaque wall; however, in this case, the slit is very narrow in one direction. This time the phenomenon of diffraction spreads the image along the direction that the slit is narrowed. The image in (b) is brightest directly behind the slit and fades out to each side. If single photons are sent one-by-one through the narrow slit, as shown in (c), eventually an image like that in (b) results. However, it is not possible to know where any individual photon will hit the screen. In (c), the image shows the seemingly random positions of impact for the first three photons sent through the slit.

To illustrate this, let's consider an example. Suppose a broad beam of light passes through a wide slit in a dark room and hits a flat screen situated a few inches behind the slit. Let's imagine the screen is constructed so that a tiny bright dot appears at each location where the screen is hit by a photon. The technical details of how we accomplish this are not important here. Let's just say we "have people for that." If the source of the light is far away, and the light beam is intense and wide enough to cover the slit, we see a sharp-edged image of the wide slit on the screen, as illustrated in Figure 4.1(a). In this case, the light beam consists of many photons that interact with the screen and the overlapping dots create a solid, bright image of the opening. Things become a bit more interesting if we make the slit extremely narrow, because light passing through a very narrow slit undergoes a well-understood wave phenomenon called diffraction that causes the light to spread out as it passes through the slit. This blurs the image of the slit on the screen. In this case, the image looks brightest directly behind the slit and fades out over some distance to each side along the narrow direction of the slit. The image no longer appears sharp and would look something like the sketch shown in Figure 4.1(b).

Now suppose we keep the slit narrow as in Figure 4.1(b) but dim the beam of light until it consists of a single photon passing through the slit at any given time. What happens now? We know a single photon will pass through the slit and hit the screen somewhere, causing a tiny bright spot at the point of impact. Quantum mechanics can tell us the probability distribution of where the photon is likely to hit the screen. That probability distribution would look just like the illumination pattern we would see with many photons, as in Figure 4.1(b). The photon is most likely to hit directly behind the slit and the probability of a strike fades off to each side. What we cannot determine, according to quantum mechanics, is the *exact location* of the strike. Each photon would appear to hit the screen at some random point within the region that would be illuminated if the light were intense.

Figure 4.1(c) shows what we might see after sending three photons through the slit, for example.

So far, so good. To see the controversy, let's follow the evolution of the photon wave function in this single photon scenario. Initially the photon is part of the beam (dim though it is) and its wave function is confined to the envelope, or path, of the light beam. As the photon passes through the slit, the wave function must spread out. We know it does this because there is some probability that the photon hits the screen off to one side and that probability is represented by the strength of the wave function at that point. The photon wave function must be extended in this fashion all the way up to the very instant that the photon hits the screen. When the photon strikes the screen, the interaction of the photon with the screen occurs at the very particular point where the bright dot appears, and the location of the photon is known with certainty. At that instant the wave function must be present only at the particular point where the dot appears, because that's where the photon is located at that moment of interaction and there is no probability that it is elsewhere. This means the wave function must make an instantaneous transition from being spread out over a finite region to being concentrated at a single point. This may not sound like a big deal, but something collapsing instantaneously over a finite distance is quite bothersome to many physicists because it seems to involve the transfer of information at a faster-than-light speed. This violates the basic sense of cause and effect that is core to our ideas about how the universe works. How is it that the wave function on the left side of the image region instantly finds out that the photon just hit the screen on the right side of the image region without violating causality?

Through much of the last century most physicists would think about an event such as our photon hitting the screen using what is called commonly the Copenhagen interpretation of quantum mechanics. This view came out of discussions between Niels Bohr and Werner Heisenberg in the late 1920s.[3] According to the Copenhagen interpretation, the wave function of the photon propagates in its extended form until it is collapsed by an interaction or observation. The faster-than-light thing should not be too bothersome because the wave function is never observed anyway. The discontinuity in the way that the system is described is real. Prior to the interaction, the full set of possibilities must be preserved—as in the wave function description—and after the interaction, all possibilities but the one that is realized are discarded.

It's all well and good to discuss tiny particles, but the absurd nature of this business is best brought out by the famous example of Schrödinger's cat. Imagine that we have a sweet fat cat and we place him in a box with a single radioactive atom, a radiation detector, and a vial of poison. The idea is that if the radioactive atom decays it will set off the radiation detector, which will cause the poison to be released and kill the cat. As long as the radioactive atom does not decay, the cat is alive and well, happily curled up in the box (no doubt sleeping).

Now there's no reason to get excited. It's just a pretend example. I love cats. In fact I have a happy kitty curled up and snoring beside me right now. I'm not sure what he does when I'm at work, but he does not seem too concerned with quantum mechanics at the moment.

Let's say that we close the cat in the box with our contraption for one hour. Also, let's assume that the radioactive atom in the box has a 50-50 chance of decaying in that hour. We can't know at any moment whether or not the atom has decayed. All we know is that at the end of the hour there is a 50-percent probability that it will have decayed. So, at the end of the hour, if we were to ask what is the quantum mechanical state of the cat just prior to opening the box, it would be a superposition of half-dead and half-alive. Only after opening the box and looking inside—acquiring more information—does the wave function of the cat collapse to being either dead or alive.

Does this mean that Schrödinger's cat exists in the box in some strange half-dead and half-alive limbo? Certainly the mathematics of quantum mechanics implies that the cat exists in such a bizarre state until we acquire the additional information that collapses the wave function. Is this superposition—this mixing of, or simultaneous existing in, the different states—real? Copenhagenists say that it makes little sense to talk about the reality of the superposition until an observation is made that collapses the wave function into a definite state—that is to say, the observation forces the wave function to change instantaneously into something that agrees with the observation. They argue that definiteness and causal connectedness are only critical for those things that we can actually measure, and the bothersome aspects of quantum mechanics are not things that can be measured.

The Copenhagenists may be wrong. Though it is beyond the scope of this book, there have been elegant experiments done that show quantum superpositions seem to be real, and these strange instantaneous transformations over a large distance (a physicist would call it a nonlocal effect) seem to happen.[4] This is an area of current investigation.

Quantum mechanics works well. The mathematics of the theory is well established and unambiguous, at least in terms of how it is used as a tool for calculations. Beneath the universally acclaimed and proven utility of the theory, the discussion has been about what it all means for our view of the universe. In the Copenhagen interpretation, a boundary is placed between the quantum world and the classical "macro" world, and we only need concern ourselves with things that we can observe in the macro world, though the location of the boundary between the two worlds is sometimes a bit fuzzy.[5] The Copenhagen interpretation of quantum mechanics has been the dominant framework for interpreting quantum mechanics for most of the last 100 years. Also, as we will see, some people have used a modified version of the Copenhagen view to develop ideas that have found a place in our multiverse taxonomy in spite of the fact that the claims are dubious scientifically. I'm referring to what I call the multiverse of wishful thinking where consciousness is part of collapsing the wave function or can even preferentially collapse the wave function. More on this later.

Other interpretations of quantum mechanics have been proposed. The most intriguing of these—and one that appears to be rapidly gaining in acceptance among physicists[6]—is the so-called "many-worlds interpretation." This theory was invented in 1956 and 1957 as the PhD project of a Princeton graduate student named Hugh Everett III.[7] When published initially by Everett, the work was entitled "Relative State Formalism of Quantum Mechanics." A decade later the idea was given the "many-worlds" moniker by Brice DeWitt, who had been editor of the scientific journal that initially published Everett's work.

To Everett, proponents of the Copenhagen interpretation of quantum mechanics were being inconsistent in how they described small things using quantum mechanics and used different—classical—rules for the macro world where things could be observed. It was this inconsistency that led to the strange effects like instantaneously collapsing wave functions. To get around those problems, Everett formulated a theory that used quantum mechanics to describe everything, small or not. What he came up with was more elegant and consistent than previous treatments of the problem, though still hampered by bothersome issues, albeit different ones from those plaguing the Copenhagenists.

In his relative state formalism theory, Everett imagines that the universe consists of one huge, unimaginably complex wave function that evolves smoothly according to the rules of quantum mechanics. Within Everett's theory, as with all quantum mechanical models, the wave function exists in what physicists and mathematicians call "Hilbert space." In the classical—real—world in which we live our everyday lives, we think of objects moving in three dimensions. We can track that motion mathematically as a function of time by looking at three numbers and how they change with time. Those numbers correspond to the position of the object in each of the three dimensions. Imagine tracking a cat moving around a grocery store.

Initially, it is on aisle 2, halfway down the aisle, on the floor. Then it moves to aisle 3, near the cash registers, up on the second shelf. Each position is specified by three numbers in the spatial coordinate system of the grocery store. If you were to track the positions of *two* cats moving independently in the grocery store, it would require six numbers: three for each cat. Physicists and mathematicians would say this problem must be analyzed in a six-dimensional space. A similar game is played in describing a wave function in quantum mechanics. To understand the problem, you must specify the things that can change and what is being analyzed. In this discussion, we talk about cat locations. In quantum mechanics it might be something similar like the location of a particle or the locations of many particles. In addition, the energy and other things that we've not discussed (spin, for example) might also need to be part of the wave function. Finally, the mathematics involves complex numbers. Physicists call this complex, potentially infinite-dimensional space in which the quantum mechanical states are specified *Hilbert space*. Two different states of the same particle are separated in Hilbert space in the same way two different locations of the same cat in the previous example can be separated in grocery store space.

According to Everett, the wave function of the universe lives in an infinite-dimensional Hilbert space. In addition, this wave function evolves over time in a fashion that preserves probability. In other words, the sum of the probabilities of all the different possibilities is always 100 percent. If you flip a coin, for example, there is a 50-percent probability of coming up with heads and a 50-percent probability of tails, but there is a 100-percent probability that you will get either heads or tails. Physicists call such a theory—where the sum of all the probabilities is always 100 percent—unitary. The wave function contains within it all possibilities of what might happen, and if we were to sum up all the probabilities of all the things that *could* happen, we would get a probability of 100 percent. Though the wave function can evolve and the probabilities of different things happening might change with time, the *total* probability stays constant.

At the time of Everett's work, there was nothing all that new about a wave function evolving in a unitary way in Hilbert space. That was standard quantum mechanics. What was new, however, was that Everett hypothesized that the wave function *keeps* evolving smoothly that way rather than collapsing in order to appear like what we observe in the classical macro world. In the Copenhagen interpretation of quantum mechanics the wave function evolves smoothly up to the point an observation is made, and then it collapses suddenly into the form of wave function dictated by the observation. According to Everett, nothing really changes when the observation is made. The overall wave function keeps on evolving smoothly. What *does* happen that's new in Everett's view is that the one classical reality splits into a superposition of many classical realities, of which we are only able to perceive one so that the appearance of wave function collapse is preserved. In other words, for each quantum mechanical process that happens, the universe splits into a series of parallel realities, each of which is as real as any other!

To emphasize the difference between the Copenhagen and Everett interpretations of quantum mechanics, let's return to visit Schrödinger's cat. Recall that our sweet, fat cat in the box was unwittingly put in a life-and-death situation at the mercy of the quantum mechanical decay of a radioactive nucleus. According to the Copenhagenists, the cat exists in a quantum mechanical super position of half-alive and half-dead until the box is opened, at which point the wave function collapses such that either a dead cat or a live cat is observed, and there is only one classical reality, or universe, if you will. According to Everett, the quantum mechanical wave function evolves such that both threads of reality continue with equal probability along different paths in the Hilbert space. Along one path the observer sees a dead cat and sadly prepares to bury it. Along the other path the observer sees a happily sleeping, live cat when he opens the box. Both universes are equally valid in this picture—thus, the appropriateness of the name many-worlds interpretation of quantum mechanics.

Everett's work was revolutionary, and it contradicted the paradigm established by the influential Bohr and others. So, perhaps it isn't surprising that Everett had difficulty finding acceptance for his ideas. The version of the thesis that was published in 1957 was heavily edited and watered down, and the original version didn't appear in print until 10 years later. One of Bohr's theoretical physics colleagues in Denmark called Everett's work "theology," which is not generally considered a complement in scientific circles.[8]

To be fair to the critics, Everett's theory left some questions unanswered. If all valid quantum mechanical states continue on in a universe, why is it that we never see macroscopic superpositions such as the strange, half-dead and half-alive cat? After all, that is a perfectly good quantum mechanical state. Another way to ask this question is why is it that we see only the classical macro states that we would expect to see if the wave function collapsed? We see the dead cat or the live cat, but why not the superposition?

These questions about—or *objections to*—the many-worlds interpretation of quantum mechanics have been addressed by the idea of quantum mechanical *decoherence* proposed by physicists Dieter Zeh, Wojciech Zurek, and others.[9] It turns out that quantum mechanical superpositions persist only so long as they are completely isolated. Even the collision of a single molecule or photon with the system is enough to cause the superposition to become less apparent to us or, as a physicist would say, *decohere*. Because this typically happens in a very, very short time, particularly for macroscopic objects, superposition states are not observable typically. Quantum mechanical superpositions have been observed and studied for microscopic states, however, where the system can be isolated.[10] These states form the basis of quantum computing, for example.

To get an idea of how decoherence works, let's go back to our poor, quantum mechanical kitty. This time, let's look at a graphical representation of the state of the cat when the box is closed. Let's imagine that the "state of the cat" is represented by an arrow, something a physicist might call a state vector. Referring to the sketch in

Figure 4.2, let's say that when the cat is alive the arrow points in the vertical direction and when the cat is dead the arrow points in the horizontal direction. When the box is closed, there's a 50-percent probability of either case and the state of the cat is represented by the arrow midway between the live and dead axes. If the cat is completely isolated from the rest of the universe, the live and dead axes are the dimensions of the Hilbert space for the quantum mechanical problem. In this case, the strange superposition state is a valid quantum mechanical state, and we should see it if the wave function does not collapse to one of the axes when an observation is made, à la Copenhagen.

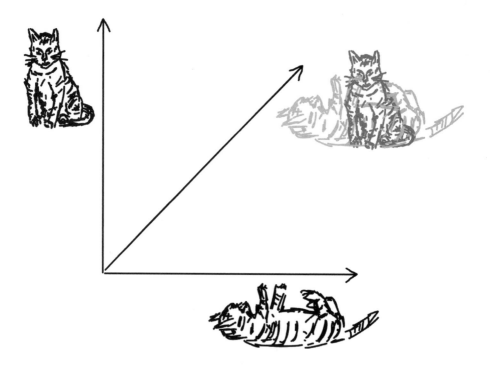

Figure 4.2: Schrodinger's cat illustrates quantum superposition.

In the many-worlds theory, the wave function *doesn't* collapse, and yet we don't see the cat in the strange superposition state. Why is this? The answer, according to Zeh and Zurek and the other decoherence gurus, is that the cat is not completely isolated. If the cat wave function interacts with anything else, like a photon or a molecule of the surrounding gas, the cat's wave function becomes *entangled* with the wave function of the other object. This increases the number of dimensions of the Hilbert space in the problem. For example, let's imagine that a molecule that can be in state A or state B interacts with the cat. Now the problem is more complex. The cat can be alive with the molecule in state A; the cat can be alive with the molecule in state B; the cat can be dead with the molecule in state A; the cat can be dead with the molecule in state B; the cat can be in a 50-50 dead/alive state with the molecule in state A, and so forth.

With the increased dimensionality, the number of possible states for the system wave function has increased. Now there are many more possibilities. But, because we require that the system evolve in a unitary fashion, where the total probability of all the states is 100 percent, there is now a smaller probability overall that the cat will be found in the strange half-dead and half-alive state where it started. That probability is smaller because there are other states that have nonzero probability and it is a zero sum game; all the probabilities always add up to 100 percent. If the cat interacts with many molecules and photons, the superposition state becomes entangled with more and more of these objects and the probability of seeing it in the original state becomes less and less. The superposition state does not collapse, but rather *bleeds* or *leaks* into the environment. In decoherence theory, the classical states where the cat is either dead or alive are more robust against this probability leakage and persist, which is why those are the states we observe.[11] No collapse necessary.

Within the modern view of the many-world theory, strange superpositions of macroscopic objects are formed as part of the universal wave function. However, within an extremely short time—far shorter than it takes to make an observation, for example—these states disappear due to interactions with the environment. What remains are the robust classical states that we observe. These states are separated within Hilbert space so that an observer is only aware of one of those states. The cat is either dead or alive, but not both. Still, in our cat example, neither the dead or alive state is preferred over the other (mathematically, that is), and both persist in the large wave function of our universe, effectively producing separate realities. To the observer in each thread of this story, it appears that the wave function collapsed leaving them with the live (or dead) cat. Each reality continues to evolve separately as part of the greater wave function.

Because we are unable to step out of our place in the infinite-dimensional quantum mechanical Hilbert space of the universe and see the great wave function evolve overall, the threads created in the many worlds theory satisfy our definition of a multiverse. In this case, the different branches delimited by decoherence in quantum processes are dimensionally separated in the quantum mechanical Hilbert space and not necessarily separated in time or regular space. Recognizing this in Appendix A, I categorized the many-worlds multiverse as a dimensionally separated multiverse. The many-worlds theory of quantum mechanics is what Max Tegmark calls the Level III multiverse.

Is the many worlds multiverse real? Well, let's look at the ingredients. The formalism of quantum mechanics works exceedingly well and is supported by a vast amount of scientific evidence. Quantum mechanical superpositions and decoherence have been observed and studied, particularly for small, isolated systems. The last ingredient of this strange mixture is the unitary evolution of the quantum mechanical wave function. According to Tegmark, "If the time-evolution of the wavefunction is unitary, then the Level III multiverse exists, so physicists have worked hard on testing this crucial assumption. So far, no departures from unitarity have been found."[12]

If the many worlds of Everett are real, it is reasonable to ask: How many worlds are there? It seems as if the number must get out of hand rather quickly. It might be, in fact, that this question really makes no sense. I am intrigued by a picture painted by Oxford quantum physicist and philosopher David Wallace, in a recent paper. Wallace says that decoherence causes the universe to split into emergent branches, but there is no way to decompose macroscopic reality into countable fine-grained histories. According to Wallace, "asking how many worlds there are is like asking how many experiences you had yesterday, or how many regrets a repentant criminal has had. It makes perfect sense to say that you had many experiences or that he had many regrets; it makes perfect sense to list the most important categories of either; but it is a non-question to ask *how many*."[13]

Although each quantum process in an Everett world causes a branching, so many things are happening at once and over so large a space, it is probably better to think of the branching in terms of a continuum than in terms of distinct worlds. A metaphor used by Wallace inspired me to the following picture. Imagine a large room where several hundred sheets of blank paper are spread out on tables. Now, suppose a crazed teenager sprays and dribbles ink drops of various sizes randomly on the sheets of paper. Then, magically and without spreading the drops, the papers are rearranged into a single, tight stack. A few of the smallest ink drops will have already dried and will be confined to a single sheet of paper in the stack, though the vast majority of drops are still liquid and bleed outward both along the sheet and into sheets above and below them. Large drops will spread ink into many, many sheets of the stack. After a few days, when the ink has all dried, imagine paging through and comparing the sheets. The ink pattern on each sheet is distinct, though very similar to the sheets above and below it. As you move further into the stack above or below a particular sheet, you find the similarity fading. If the sheets of paper are very large and very thin, the distinction between adjoining sheets is blurred; yet, clearly, a given sheet is distinct from sheets far away in the stack.

To tie our strangely inked sheets to the many-worlds picture, imagine that each sheet of paper represents a complete Everett world or branch. In this analogy, if the sheets are stacked after the ink dries, they are quite independent of each other. This is the vision of the non-interacting branches that often comes to mind when you initially confront the many-worlds picture. But, if you imagine one quantum process happening that, through decoherence, causes a branching of the universe in Hilbert space, the two branches are almost identical immediately after the branching. For example, the branches may differ in the cat being alive or dead, but the rest of the universe in the two branches is identical, at least for an instant (if you'll allow me to use the classical concept of simultaneity that ignores the complications of relativity). The similarity fades as new branches emerge from the initial branches over time. In our ink-blotted stack of worlds, the similarity between the worlds is represented by the similarity in the ink patterns. Sheets close together in the stack are very, very similar, but the similarities fade as the distance between the sheets increases. The change is more or less continuous as you move through the stack. As in many worlds, this continuum makes the question of "how many distinct universes?" nonsensical.

An essential element of the many-worlds picture is the fact that within the emergent streams of reality, we are aware of only one. It has been postulated that it is not so much that there is a splitting of the universe due to decoherence, but rather it is only our awareness—or our mind—that splits. This idea is known as the many-minds theory.[14] According to Albert and Loewer in an introduction to the many minds view, "What we want is an 'interpretation' which explains how it is that we always 'see' macroscopic objects as not being in superpositions and never experience ourselves in such a superposition."[15] In other words, the scheme is designed to avoid the wave function collapse and explain why we do not perceive the strange superposition states that should accompany the continually evolving wave function.

As far as I can tell, the many-worlds theory and the many-minds theory really differ in ways that would interest a philosopher more than a physicist. In both cases the fully unitary evolution of the universal wave function happens. In many worlds, different streams of reality evolve separated in Hilbert space and there is no way for an observer in one stream to perceive what is happening in a different stream. In the many minds view, the splitting is one of our perception. We are only able to perceive particular streams that correspond to the classically accepted potential realities that are the emergent streams in many worlds. It isn't clear to me how the continuous splitting of minds is associated with the physical reality through time because different mind states could lead to different physical actions or paths in time. I might be missing something, but many minds seems to be a complicated way to avoid perceiving physical superposition states—something that decoherence handles quite nicely anyway.

The connection between quantum mechanics and consciousness has been widely discussed through the years by scientists and philosophers such as Euan Squires, Nick Herbert, and Roger Penrose.[16] Victor Stenger has written a very nice review of the topic.[17] I will not attempt to do justice to this subject here.

There is a school of thought that consciousness essentially creates reality. I think this comes, in part, from the idea—closely associated with the Copenhagen view—that wave function collapse is caused by observation. As you might imagine a number of New Age thinkers and personal empowerment gurus have jumped on this idea of "mind controls reality." It is advocated shamelessly in the movie *What the Bleep Do We Know!?*, which was released in 2004. This movie consists of numerous interview clips of New Age and personal empowerment personalities, as well as a few scientists and a theologian. Though some of the clips seem as if they are taken out of context, the message is clear: Your mind selects among the various quantum possibilities, essentially causing the reality we perceive.

For example, in this movie Fred Alan Wolf, a physics PhD known for touting a connection between quantum physics and consciousness, states, "There is no out there out there that is independent of what's going on in here."[18] In another clip, Amit Goswami, a retired professor of theoretical physics, says, "Quantum physics calculates only possibilities. But, if we accept this, then the question immediately comes 'What chooses among the possibilities to bring the natural event of experience?' So, we directly, immediately, see that consciousness must be involved. The observer cannot be ignored."[19] At another point in the movie, the narrator states, "We've been conditioned to believe the external world is more real than the internal world. This new model of science is just the opposite. It says what is happening within us will create what is happening outside of us."[20]

The idea that external reality can be controlled consciously has been widely publicized in the phenomenal best-selling book by Rhonda Byrne entitled *The Secret*.[21] In this book, Ms. Byrne pushes the so-called "law of attraction," which is introduced as "Everything that's coming into your life you are attracting into your life. And it's attracted to you by virtue of the images you're holding in your mind. It's what you're thinking. Whatever is going on in your mind you are attracting to you."[22]

Though the law of attraction was bouncing around long before quantum mechanics was invented, Ms. Byrne claims that it is rooted soundly in the laws of quantum mechanics. She says, "The law of attraction is the law of creation. Quantum physicists tell us that the entire Universe emerged from thought! You create your life through your thoughts and the law of attraction, and every single person does the same."[23] She quotes Fred Alan Wolf for support. In the book, Wolf says, "Quantum physics really begins to point to this discovery. It says you can't have a Universe without mind entering into it, and that the mind is actually shaping the very thing that is being perceived."[24] Another person with formal physics training and who is well-known in New Age circles, John Hagelin, is quoted saying, "Quantum mechanics confirms it. That the Universe essentially emerges from thought and all this matter around us is just precipitated thought."[25]

In *The Secret*, the connection between the law of attraction and quantum physics is treated as an accepted fact without the support of scientific evidence. In *What the Bleep Do We Know!?*, the idea that the mind plays an essential role in creating reality is, again, stated as evident from modern quantum physics with no supporting scientific evidence. If Hagelin, Wolf, and Goswami are making scientific statements—which seems to be the claim—they are referring to work I've not seen or heard about in serious scientific circles. What they say does not follow from the science I know and it makes little sense to me. Take Goswami, for example.

He currently calls himself a "quantum activist" and is pushing a "paradigm based on the primacy of consciousness."[26] On the main page of Dr. Goswami's Website is a section that reads:

> More than just theory, quantum activism is the moral compass of quantum physics that helps us to actually transform our lives and society. So let's walk our talk, and make brain circuits of positive emotions. We just do it. We practice. Let some of us be good, do good. Be with God some of the time, be in the ego some of the time, and let the dance generate creative acts of transformation. With this resolution, with this objective in mind, I invite you to become Quantum Activists.[27]

Perhaps I'm just being dense, but Dr. Goswami lost me around the "moral compass" bit. Certainly, nothing here looks familiar to me from my years of physics training, teaching, and research. In fact, in Dr. Goswami's Website, as well as in *The Secret* and *What the Bleep Do We Know!?*, I find the evidence for a connection between quantum mechanics and consciousness strikingly absent—so much so, in fact, that I'm left to infer a logical structure around what I *think* they mean. In quantum physics, it *is* established that the observer and system are not really separate. In the Copenhagen interpretation of quantum mechanics the act of observing is seen to collapse the wave function. In the many worlds plus decoherence picture the observer and observed system are both part of a universal wave function. At the extreme of the Copenhagen point of view, it is the act of observation that is important and the external reality prior to observation is murky and irrelevant because only things that can be observed are real in any way that is meaningful. In the many worlds plus decoherence picture all the possibilities evolve as part of the universal wave function and decoherence insures that strange superposition states don't last long enough to be bothersome.

The real issue, it seems, is centered in the question of to what extent the mind, through conscious or unconscious thought, can affect a quantum system. In my opinion there is no room for conscious thought to be involved with quantum processes at that level. Besides the lack of any evidence (to my knowledge) and the murky nature of the fundamental mechanism through which it could happen, the sheer number of quantum processes that take place to define our reality makes the idea that we consciously cause them unfathomable. Also, the decoherence time for quantum processes in the brain has been shown by Tegmark to be on the order of 10^{-13} to 10^{-20} seconds. That's vastly shorter than even the most fickle individual can have a flighty thought. Said more academically, it is substantially shorter than the relevant timescale over which brain processes occur.[28]

Could reality be a product of the unconscious mind? Perhaps I should leave this one to the philosophers, but it's hard to stand on the sidelines completely. We know that quantum processes happen. We know that quantum superpositions exist and that a typical decoherence timescale in the brain is 10^{-13} to 10^{-20} seconds. What do we define as "the mind"? If we consider the mind to be at the mercy of external reality—that is to say, determined by physical processes in the brain—then, as Tegmark has shown, the timescale over which quantum choices are made is much shorter than the timescale of the relevant processes in the brain. In that case, the mind really does not have the opportunity or capability to affect quantum processes and there is no scientific evidence to support the claims made in *The Secret* and *What the Bleep Do we Know!?*.

If we suppose the mind is independent of external reality, my arguments against the "reality springs from the mind" idea are not valid. This is the case where the mind is independent of the body and all the physical constraints that run the system to the best of our knowledge. If the mind were truly independent of matter and forces as we understand them, it makes little sense to discuss a connection between the mind and quantum mechanics. As far as I know, in spite of the fact that it is a constant thread in *Star Trek* episodes and New Age literature, there is no scientific evidence to support the concept of the mind being independent of the physical world.

Don't get me wrong. I'm sure the mind has an effect on the individual human condition. I think the power of positive thinking is very real. It is well established that one's state of mind can affect health.[29] Anyone who has played or watched much in the way of sport will attest to the fact that the mental game is an important element—not just knowing what play to make, but to believe that you can make it. I think it's important to be in a frame of mind that is receptive to achieving one's goals. All these things are well and good, but they don't follow from quantum mechanics any more than does the surgeon taking a scalpel to a cancerous tumor. All people—you, me, and the surgeon—the balls, the scalpel, and so forth, are all made of subatomic particles that are subject to the laws of quantum mechanics. Still, even the most arrogant physicist has to admit that we cannot understand the emergent properties of complicated things like basketball players and heart surgeons through our current scientific theories—including quantum mechanics.

In spite of the lack of solid scientific evidence, the popularity of *The Secret* shows that many people are intrigued by the idea that the conscious mind can play a role in selecting among potential realities. If we combine this idea with the emergent threads of the many-worlds theory, we have what I call the "multiverse of wishful thinking." This is a multiverse where a person's mind and beliefs can guide a person among the various Everett worlds that are continually evolving in the universal wave function. I classify this as a faith-based multiverse. It is something in which a person can choose to believe, but it is unsupported scientifically.

Returning to quantum mechanics, another place where the theory plays a critical role in physics is in our understanding of the nature of forces and the fundamental particles in the universe. Because these forces and particles mold our universe and tell us its history, it is important that we survey what is known about them before we discuss the many cosmological multiverse concepts. That's our next stop.

Now, if you'll excuse me, it's time for me to meet with my life coach. We are practicing my visualization of the universe in which hoards of people are lining up to buy this book!

5
Of Boxer Shorts and Charmed Quarks

ack in the day, when I was a kid roaming the nearby creeks and fields after school, I'd swing by my home every now and then for a snack. Sometimes my buddies would join me, and we'd deposit a few muddy footprints by the door and raid the kitchen. That's the way it worked. Some days we'd trash my house. Some days it was someone else's house. Anyway, when my house was the target, I always had to ask my friends to wait just outside the front door for a minute. Then I'd run into the house and warn my dad that we had company so that he'd get dressed and avoid an awkward event. You see, my father always sat around the house in boxer shorts and nothing else. It only took a couple of red-faced friends for me to learn to be religious about checking inside the house before barging in with a buddy.

Does this sound strange to you? When I was growing up I thought it was normal. I figured everybody's dad sat around the house in their boxers. I always assumed that someday the urge would hit me, too. Suddenly one day, I figured, I would wake up overwhelmed by the need to lounge around my castle in my underwear. Alas, it's never happened—no doubt, much to the relief of *my* kids, who would be rather horrified if they waltzed into the living room and found me sitting around like my dad.

I suppose anyone looking back over the years at their youth would find a few things that seem strange. In fact, because "normal" is more or less what seems familiar, I suspect that if you selected any stranger at random and dug deep enough into his or her behavior you would find something that strikes you as very odd.

To me, quantum mechanics has always seemed like people in this respect. When you first learn a little about quantum mechanics, it seems a bit strange. But, once you've been exposed to the field and the techniques for a time, you sort of get used to quantum mechanics. You go into the "shut up and calculate" mode or, better yet, the "shut up and hope someone else has already calculated it" mode. Familiarity numbs you to the aspects of quantum mechanics that seemed so very strange initially.

The things that make quantum mechanics seem so odd to most of us are the non-locality and indeterminism characteristic of the theory. It's not simply a matter of human discomfort. These qualities of quantum mechanics truly are limiting. Up until the discovery of quantum mechanics, physicists felt confident that, in principle, they could measure arbitrarily well the position and momentum of all the particles in a system of interacting bodies and, armed with the knowledge of the forces between them, predict the future of the system. That system might even be very large—say, something like the universe. But, quantum mechanics, as manifest in the uncertainty principle, tells us it is not possible to know both the position and the momentum of a particle infinitely well. Consequently, any prediction of the future has inherent uncertainty built into it regardless of the computing power at our disposal. Because this uncertainty is integral to quantum mechanics, it cannot be overcome even with improved measurements.

As sad and frustrating as it seems to lose the power of determinism, there's no need to shed tears. In fact, it turns out that the uncertainty inherent in quantum mechanics is incredibly liberating and has driven much of the progress in fundamental physics in the last century.

The key to the liberating side of quantum mechanics is a different form of Heisenberg's uncertainty principle, one that relates uncertainties in the complementary variables of energy and time. To see how this works, imagine a particle that exists for a very short time. Because the lifetime is very short, the uncertainty in the lifetime is very small—even if the uncertainty in the lifetime is as long as the lifetime itself. According to the uncertainty principle, this means we cannot know the energy of the particle very well. This is tantamount to a license to (seemingly) break energy conservation as long as we do it over a very short time. This little apparent loophole in energy conservation provides a strange, Harry Potterish avenue to an understanding of the forces of nature via a theoretical framework known as quantum field theory.

How is it that an uncertainty in the energy of something can lead to a force? Imagine an electron moving through space. The uncertainty principle says that this electron could emit other particles like photons and seemingly break the conservation of energy as long as it happens over a very short time. It's not a *true* breaking of energy conservation unless the uncertainty principle is violated. Still, outside of the need to stay within the confines of the uncertainty principle and the preservation of a few other things like total momentum and the total electric charge, almost anything could happen. For example, the emitted photon could turn into an electron and positron pair, and the electron or positron might emit a photon, and so forth. Many different possibilities could happen, a few of which are shown in Figure 5.1 on the following page. As long as the chaotic mess only sticks around for a time short enough that the uncertainty principle is satisfied, things are okay.

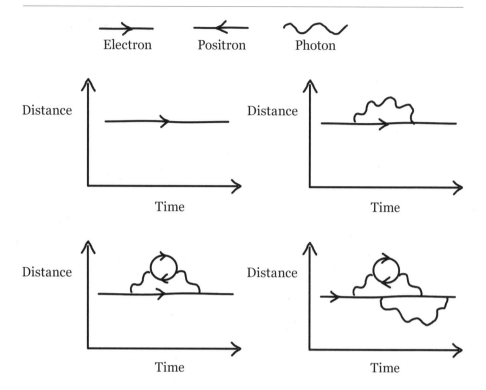

Figure 5.1: Feynman diagrams illustrating a few of the quantum possibilities that must be considered when thinking about an electron traveling in a vacuum in quantum field theory.

The diagrams in Figure 5.1 are examples of Feynman diagrams, named after their inventor, Richard Feynman—the very same Feynman whose work inspired Tegmark to turn from economics to physics. Feynman was one of the great characters of modern physics. He was a bongo player and a jokester and a good story-teller, as well as a world-class theoretical physicist. He was awarded the Nobel Prize in Physics in 1965 along with Sin-Itiro Tomonaga and Julian Schwinger for "their fundamental work in quantum electrodynamics, with deep-ploughing consequences for the physics of elementary particles."[1]

Feynman had a sharp wit and a way with words. He published a number of popular, entertaining books and audio recordings relating his philosophy and stories from his experience. One of my favorite quotes attributed widely to Feynman is "Physics is like sex: sure, it may give some practical results, but that's not why we do it."

Feynman once told a group of science teachers, "Science is the belief in the ignorance of experts."[2] This is a saying he took to heart, claiming to always calculate things for himself. He showed the world the effectiveness of this investigative approach during a televised congressional hearing in 1986. The hearing was held as part of the process to determine the cause of the Space Shuttle *Challenger* accident in 1986. On January 23rd of that year, the *Challenger* lifted off under clear skies in Florida and exploded 73 seconds into the flight with seven astronauts on board. It was a national tragedy. What ensued was much speculation and finger-pointing. The NASA shuttle fleet was grounded and President Reagan appointed a commission to look into the accident and find the cause. Richard Feynman was part of that presidential commission, usually called the Rogers Commission after William Rogers, who was the chair of the committee. In the course of the investigation, evidence pointed to a failure in an O-ring seal situated between segments of one of the *Challenger*'s solid rocket boosters. The job of the O-ring was to keep the hot gases inside the rocket from escaping between the booster segments. The *Challenger* was launched the morning after a night with temperatures below freezing, and one of the issues became the performance of the O-rings at cold temperatures. During the commission's questioning of representatives of Morton Thiokol, the contractor responsible for the construction and maintenance of the solid rocket boosters, Feynman laid the question to rest in a fashion that was, well, typically Feynman:

> took this stuff that I got outta your seal and I put it in ice water. And I discovered that when you put some pressure on it for a while and then undo it, it doesn't stretch back. It stays the same dimension. In other words, for a few seconds at least—more seconds than that—there's no resilience in this particular material when it's at a temperature of 32 degrees. I believe that has some significance for our problem.[3]

Feynman diagrams are also typically Feynman. They are a powerful way to visualize and calculate all the quantum possibilities for a particle interaction. Feynman diagrams are two-dimensional pictograms of particle interactions. In these diagrams, one axis represents time, and the other represents position along one dimension. A particle at rest would be represented by a straight line parallel to the time axis. These diagrams are more than pretty pictures. They are drawn according to rules, and each part of the diagram represents some of the mathematics involved in a calculation. That said, we'll look at them as pretty pictures of things that could happen.

As Figure 5.1 shows, there are many possibilities to consider even for something as simple as a particle moving through a vacuum. So, it is not appropriate in the world of quantum mechanics to consider an electron as something like a marble. Instead the electron is described as a superposition, or a simultaneous mathematical mix, of all the possible quantum states allowed by the uncertainty principle (and other constraints such as electric charge conservation). So, it is more appropriate to think of the electron as a seething cloud of particles appearing and vanishing so quickly that they cannot be observed. These transient states are called "virtual particles." Though they can't hang around very long without violating the uncertainty principle, the virtual particles carry the momentum and electric charge of the electron—which is the essence of the electron. Because virtual particles carry momentum, things get interesting when two electrons pass near each other and exchange a virtual particle. An exchange of momentum is what we know as a force. This exchange of virtual particles, and the momentum they carry, is the fundamental concept of a force in modern quantum field theory.

This so-called "exchange force" idea isn't as crazy as it may sound at first. Imagine two ice skaters standing a few feet apart on a frozen pond. Suppose one of the skaters holds a heavy ball and tosses it to the other skater. The act of tossing the ball on one end and catching it on the other will cause each skater to move away from the other. The act of exchanging the ball produces a force between the two skaters.

In quantum field theory the exact nature of the force depends on the type of virtual particle that is exchanged. To date, physicists have identified four fundamental forces of nature: gravitation, electromagnetism, and the strong and weak nuclear forces. Three of these four forces—electromagnetism and the strong and weak nuclear forces—are quite well understood in the context of the quantum field theoretical framework known as the Standard Model. In electromagnetic interactions, the exchanged particle is a photon; the strong nuclear force between particles is conveyed by a particle called the gluon; and the weak nuclear force is carried by particles called the Z and the W.

As of yet, there is no known quantum theory of gravitation that really works, as will be discussed in subsequent chapters. Fortunately, the general theory of relativity provides an excellent description of gravitation on all but the highest energy and shortest distance scales, and is sufficient for most things. When quantum gravitational effects are expected to be important, physicists rely on educated, model-specific speculation to fill in the blanks—at least for now.

Of course it's not possible to separate the forces from the particles that do the interacting, and, throughout the years, physicists have discovered quite a zoo of particles. Typically, it seems that when scientists are faced with a chaotic mess like this zoo, they seek patterns and categories, and make up crazy names to distinguish the different beasts. This strategy is a very important part of the scientific method and has brought us plant and animal taxonomy in biology, stellar types in astronomy, and the periodic chart and chemical nomenclature in chemistry, among other things. Of course, particle physicists were determined not to be outdone by the astronomers, biologists, and chemists. They rose to the occasion, and named and categorized and classified and organized and eventually made sense of the mess.

Particles in the Standard Model

Quarks		Leptons	
up	u	electron	e
down	d	muon	μ
strange	s	tau	τ
charm	c	electron neutrino	v_e
bottom	b	muon neutrino	v_μ
top	t	tau neutrino	v_τ

Gauge bosons		Higgs boson	h
photon	γ		
W and Z	$W^{+/-}, Z^o$	Hadrons (not fundamental)	
gluon	g	baryon	qqq
		meson	$q\bar{q}$

Figure 5.2: In the Standard Model, the quarks, leptons, and gauge bosons are fundamental particles. The gauge bosons convey the force between other particles: The photon carries the electromagnetic force; the W and Z convey the weak force; and the gluon is responsible for the strong force. Baryons and mesons are not fundamental. Rather they are made of three quarks or a quark and an antiquark, respectively.

According to the Standard Model there are three primary categories of fundamental particles: quarks, leptons, and gauge bosons. The basic particles in the Standard Model are shown in Figure 5.2. Quarks carry fractional charge and can combine in twos and threes to create heavier particles such as protons, neutrons, and many other particles that are less commonly known, such as pions and kaons and B-mesons, among others. There are six distinct types of quarks, segregated by what the particle physicists call flavor. Yes, I'm serious: *Flavor* is a real scientific term. The quark flavors are, in order of increasing mass, up, down, strange, charm, bottom, and top. Yep, I'm still being serious.

The most famous of the leptons is the negatively charged electron that is found in the outer reaches of atoms. The electron is joined in the lepton category by two heavy cousins, the muon and tau, which are identical to the electron in every way except they are 200 and 3,500 times heavier, respectively. The lepton sector is completed by three very light, electrically neutral particles known as neutrinos.

Finally, the gauge bosons are the particles that mediate the forces. If a photon is exchanged, for example, the interaction is electromagnetic. This type of interaction can only occur between electrically charged particles. Gauge bosons called gluons mediate the strong force, while the W and Z particles carry the weak force.

In addition to the zoo of particles, it turns out that nature has supplied us with an equivalent zoo of antiparticles. An antiparticle is identical to its particle partner except that it has the opposite electric charge. The positron, for example, the antiparticle of the electron, was discovered in 1932. A positron is exactly like an electron except that it possesses a positive electric charge.

A question that begs asking is this: If matter and antimatter are identical in every way except for electric charge, why is it that the universe is made of matter rather than antimatter or a mix of the two? This is one of the most profound puzzles in physics today. We have discovered that matter and antimatter are slightly different in terms of how they interact via the weak nuclear force. Physicists are seeking to determine whether this and other potential differences are enough to create the matter asymmetry that we see in the universe today.[4]

The existence of antimatter, along with the uncertainty principle's strange loophole in energy conservation, leads to the very bizarre fact that "nothin' ain't nothin'" anymore. If we consider a bit of empty space—what a physicist would call the vacuum—we might naturally think of it as having no energy. After all, it's empty! How can it have energy? Well, things are different in the quantum world. If we consider this bit of space over a very short time, our knowledge of the energy is very uncertain. In fact, the energy in that bit of space might not be zero. The energy could fluctuate away from zero and, as long as that happens over a time scale short enough that the uncertainty principle is satisfied, it's okay. Physicists call this a "quantum fluctuation" in the energy. If that fluctuation is large enough, it can take the form of a particle-antiparticle pair (remember mass-energy equivalence, $E=mc^2$). For example, an electron-positron pair can pop out of the vacuum for the briefest of moments. Because the particles are produced in particle-antiparticle pairs where the two particles carry equal and opposite electric charge, electric charge is conserved in the process. We start with zero total electric charge and it stays that way even when the pair pops out of the vacuum. At any rate, as strange as it seems, according to quantum field theory, it is misleading to consider the vacuum as nothing. In fact, it is more accurate to consider the vacuum as a seething sea of particle-antiparticle pairs and quantum mechanical possibilities.

As wacky as it sounds, this idea that the vacuum is *something* rather than *nothing* is central to quantum field theory. Besides the amazing success of the Standard Model in general, there is strong evidence to support the idea of quantum fluctuations in the vacuum. I've little doubt that friends of mine who are theoretical physicists would gush with examples of evidence for quantum fluctuations. My favorite bit of evidence, however, is something called vacuum polarization. According to this quantum field theoretical picture I've been painting, you can imagine the electron being surrounded by virtual photons and electron-positron pairs. The electron has a negative electric charge that will attract the positrons and repel the electrons that make up the virtual pairs. So, there is a slight tendency for the pairs to be oriented with the positron closer to the center than the electron. Physicists refer to this orienting of the pairs as polarization, and so the effect overall is called "vacuum polarization." This effect slightly screens and reduces the electric field produced by the central electron. Experiments at high energy accelerators observe this effect and must take vacuum polarization into account in when they make precise measurements of other things.

The Standard Model is the current paradigm of fundamental physics. It's far more than a list of particles and forces. The Standard Model describes the constituents of matter, the forces that act between them, and the gory details of how exactly these beasts interact with each other. The concepts and techniques used in the theory evolved over a period of decades. Though the development of the Standard Model and the discoveries that engendered that development are near and dear to my heart, I will not attempt to do justice to the subject. The interested reader is referred to a Website and a couple of fun books that provide historical detail and many colorful anecdotes about this great scientific adventure.[5]

Still, it's important to take a quick look at the development of the Standard Model because it will help set the stage for things to come in the subsequent chapters as the particles and forces are intimately connected with the evolution of our universe and/or multiverse. Of course, in setting out on such an abbreviated journey, it's hard to know where to begin. Today people grow up knowing of things that were unimagined only a scant hundred years ago. What were recently amazing feats of discovery are now old news.

A few years ago I discovered that my daughter was unimpressed with my research aimed at understanding the structure of matter. She rolled her eyes, looked at me with a tinge of pity, and said, "Oh, Daddy. We already know all about *that*. I learned about it in school. Matter is made of *atoms*."

I could tell that she had concerns that I was wasting my time trying to discover things that every third-grader knows. I must have looked a little upset, because she added, "Oh, don't worry, Daddy. Maybe you should just look it up on the Web. To make you feel better, I'll tell you a joke, okay?" And she did:

Two atoms were walking down the road. Suddenly, Atom One turns around in obvious panic.

Alarmed, Atom Two says, "Dude! What's wrong?"

Atom One replies, "Man, I lost an electron back there!"

"Are you sure?"

"Dude! I'm *positive*."

A little while later the two atoms are walking down another road when Atom One turns to Atom Two and says, "Actually, I think it was a neutron I lost."

Atom Two replies, "Oh, don't worry then. You can get a new one free of charge."

Okay. Maybe she's not headed for a career as a standup comic. But, what struck me is that these jokes, referring as they do to electrons and neutrons and charge, entail a fairly sophisticated understanding of the structure of matter. Of course, she's my kid. So, perhaps I'm overly impressed with her. After all, there's nothing unusual about kids who profess to know everything. Surely throughout history kids have known it all, including whatever was the lore at the time concerning the nature of matter. No doubt Plato heard about it from his kids. "Oh Daddy! I know *all about* how everything is made of earth, fire, air, and water. Can we go to the javelin-throwing contest now?"

Though it's always true that kids know everything, or at least think they do, modern children really *are* set apart from all the kids that came before them. After all, it's only in the last 125 years or so that humankind has developed a self-consistent, scientifically sound view of matter that stands up to an amazing array of precise experiments, even in realms far removed from everyday experience. These days most children are taught a concept of matter that is backed up by scientific evidence. Never before has that been true.

The quest to understand the essence of matter has a long and noble history. There have been many outstanding successes during the last century. Still, there remain many puzzles about the nature of matter. In fact, the search for answers has become more complex and more exciting than Plato could have ever imagined. The innocent task of seeking the fundamental building blocks of matter has become intimately entwined with questions about the nature of forces and the evolution of the universe over time.

For me, the journey to understand matter begins with a few basic questions: Is there a fundamental building block out of which all matter is constructed? If so, how do we get the rich variety we see in the world around us? How are different bits of matter held together? What forces govern the behavior of matter? How does matter behave when it becomes very hot or very cold or very compressed? And on and on. You get the idea.

Strangely enough, the answers we seek depend on the size of the object being considered. It seems the perceived building block of matter is in the eye of the beholder. Historically, we have found seemingly fundamental building blocks of matter, only to discover later that they have constituents once we developed the ability to look inside them. Matter is made of atoms that are made of electrons and atomic nuclei; atomic nuclei are made of protons and neutrons; protons and neutrons are made of quarks. In a sense, the search for the nature of matter is analogous to peeling away successive layers of an onion, where the shape of each layer is determined to some extent by the layers beneath, and each successive layer is smaller.[6] Perhaps we will find a core of truly fundamental particles at the center that are the building blocks of nature, or perhaps we will continue to find smaller and smaller structures as our technology enables us to probe to smaller and smaller distance scales, peeling away layer after layer of our onion.

At any rate, according to the Standard Model, the fundamental constituents of matter—the innermost layer of the onion—are the quarks, leptons, and gauge bosons. For scale, a fine human hair has a width of 20 millionths of a meter,[7] whereas the simplest atom—the hydrogen atom—is around 0.0000000001 meters, or a tenth of a billionth of a meter[8] in size, approximately 100,000 times smaller. The smallest atomic nucleus—the proton at the core of the hydrogen atom—is 100,000 times smaller again, or 0.000000000000001 meters (10^{-15} meters). We know the proton is made of quarks. To date we have not observed quarks or the electron to have structure. They appear to be point-like to the smallest distances we've managed to probe, which is approximately 10^{-18} meters.[9]

These numbers are laden with zeros—so many, in fact, that it is easy to lose track of what they really mean. To help with this, imagine a fine human hair. If the thickness of that hair were the size of the Earth, an atom in that hair would be the size of about three-quarters of a football field, and the size of a proton in the nucleus of that atom would be approximately as wide as the thickness of six sheets of typical office paper.

As you might imagine, it's not so easy to study things that are so very tiny. To do this, physicists make use of what I call the "auto crash reconstruction technique." Suppose that you wanted to figure out how a car works but were unable, for some reason, to open the hood and inspect the engine. If you had unlimited resources, you could buy a few hundred cars, hire drivers of limited intelligence, have them drive the cars around a racetrack, and crash them into each other head-on. You could keep track of what comes out of each collision, knowing the type of cars involved and the speed of the collision. The higher the speed of the collision, the more stuff you'd see flying out. Eventually, after studying and classifying the various parts coming from these collisions, you might be able to figure out how the car engine works—having never opened the hood to peer inside. Also, by carefully examining the debris and the marks on the roadway, you could reconstruct what happened during any particular collision. For example, skid marks would indicate if one of the drivers swerved or braked just before the collision. The degree of destruction and the extent to which the parts were scattered could be used to estimate the speed of the cars prior to the collision. The angle with which the cars hit could be determined by the final positions of the cars and the various parts. There's nothing new in this. Police investigators reconstruct auto accidents regularly from the evidence on and around the roadway after the fact.

Physicists basically play this crash game with particles and high-energy accelerators. They accelerate the particles and collide them in the midst of complex detectors that are able to track and identify most of the outgoing particle debris. From this information, they can infer a great deal about what happened during the collision.

In the early years, of course, there were no manmade particle accelerators. Instead, physicists took advantage of the fact that Earth is under constant bombardment of energetic particles from space called "cosmic rays." Though the name conjures up a science-fiction film from the 1950s, they aren't all that exotic by today's standards. The majority of cosmic rays are ordinary protons, though there is a mix of other particles as well. Cosmic rays come to us in a range of energies and are produced from a variety of processes near the sun and far out in space. The Earth's atmosphere and magnetic field shield us from most of the cosmic rays. However, there is a flux of particles that reach Earth's surface from the showers of particles formed by the interaction of cosmic rays with atoms in the atmosphere. These particles provided a convenient source of collisions for physicists to study. Well, perhaps *convenient* is a strong word, because the physicists doing this business spent a good deal of time dragging detectors up mountains because there are more cosmic rays at higher elevations.

A number of different technologies have been utilized to observe particles. Generally, detection relies on particles interacting with matter electromagnetically. For example, charged particles ionize atoms (that is, knock electrons out of atoms) as they pass through material, and that freed charge can be collected and measured in a variety of ways. In the 1930s, when the study of cosmic rays was in vogue,[10] the detectors of choice were cloud chambers and photographic emulsion.

A cloud chamber is a box of air laden with water or alcohol vapor kept at a temperature where the vapor is just on the edge of condensing. In such a super-cooled state, condensation will form along the track of ionization trailing any charged particles passing through the chamber. Photographs taken of the wispy tracks before they disappear back into the gaseous state provide a faithful representation of the track path.

In a photographic emulsion detector, the particles traverse stacks of layers of photographic emulsion, the active ingredient in film, exposing the emulsion where they pass. Later painstaking analysis is required to reconstruct each event.

Cloud chambers and emulsion and other early particle-detection technologies required detailed analysis of individual events by humans. In addition to the painstaking and labor-intensive nature of the work, it was slow. So, through the years, physicists have developed detectors capable of immediately converting the ionization or light produced by the passage of a particle into an electronic signal suitable for long-term storage and analysis by a computer.

Oops! Sorry for the digression. I just think this stuff is cool.

Particle physics started in earnest in the 1930s with Carl Anderson's discoveries of the positron and the muon using a cloud chamber. The positron was the first example of antimatter to be observed. Though exciting—and worthy of the Nobel Prize in Physics— the positron had been predicted to exist by British physicist P.A.M. Dirac several years earlier.[11] The muon, on the other hand, was wholly unexpected. Initially, the muon was thought to be the particle predicted to exist by Hideki Yukawa in 1935. Yukawa's expectation was based on the fact that some sort of particle exchange should occur to bind protons and neutrons inside a nucleus and that the exchanged particle should have a mass commensurate with the uncertainty principle and the size of the nucleus, which was taken as the range of the nuclear force. The muon had some of the characteristics expected for Yukawa's particle, but it didn't interact strongly with neutrons and protons as Yukawa's particle was expected to do. In fact, the muon was discovered to be little more than a fat electron. Though distinct from the electron and 200 times more massive, it behaves exactly like an electron (once the effects due to the different mass are factored out). There was no obvious need for a particle like the muon in nature and physicists found its mere existence a profound puzzle. One of the most influential physicists of the day, I.I. Rabi, famously remarked about the muon, "Who ordered that?"[12]

Rabi is worth more of a mention here. He won the 1944 Nobel Prize in Physics for his work on the magnetic properties of nuclei in the 1930s.[13] He was involved in both the development of radar and the nuclear bomb during World War II. He also chaired Columbia University's physics department during the legendary period in the 1940s when it would have been quite difficult to walk down the hall without bumping into a current or future Nobel laureate.

The thought of Rabi always makes me shudder. When I was a beginning graduate student at Columbia near the end of Rabi's life, the great man gave several seminars to the physics department. Each time he would declare, with great seriousness, that the reason Columbia's physics graduates were so great during the 1940s and 1950s was that they were required to take and pass the dreaded, multiple-day, comprehensive written and oral graduate exams on all of physics not once, but *twice*: once to qualify to do PhD research, which was typical for graduate school, and then again after completing all the other requirements for the degree—including the thesis defense! Those comprehensive exams are a very painful and stressful experience for most students, and the prospect that I and the other students might be subjected to the ordeal two times was too horrible to imagine. Fortunately for us, Rabi's pleas went unheeded. Could the department really have been so cruel back in the 1940s? I'm not sure. I never did ask around for confirmation for fear of promoting further discussion on the matter.

The muon was hardly the only surprise for physicists as they did experiments using accelerators of ever increasing energy and detectors capable of handling the resulting collisions. They discovered and characterized many new particles in the several decades following World War II—some expected, some not. The pion, the particle predicted by Yukawa in 1935, was discovered in 1947. That year also saw the first surprising examples of particles whose behavior was so anomalous as compared to the other particles seen up to that point that they were called strange particles.[14]

By the early 1960s the number of known subatomic particles had grown to more than 75, and they possessed a seemingly random jumble of properties. Fortunately, in 1964 Murray Gell-Mann at Caltech and George Zweig at CERN independently proposed many of these particles were not fundamental, but rather constructed from three other particles, each with a fractional electric charge. Gell-Mann called these particles "quarks"; Zweig referred to them as "aces." Gell-Mann's terminology won in the end, perhaps in part because Zweig had enormous difficulty getting his paper published. Whatever they were to be called, the reality of these beasts was far from clear initially, though the theoretical simplicity of being able to construct the many from the few was appreciated by the physics community.

Initially Gell-Mann called the three odd little fractionally charged particles that he proposed in his model "quorks," which was a whimsical word he'd used previously meaning something like "odd little things." About this time, he read *Finnegans Wake* by James Joyce and ran across a passage beginning as:

Three quarks for Muster Mark!
Sure he hasn't got much of a bark,
And sure any he has it's all beside the mark....

With that, Gell-Mann's mythical trio of particles became "quarks."[15]

According to the Gell-Mann's scheme, the three quarks were named up, down, and strange. Though it may be hard to believe, there was some reasoning behind these weird names. "Up" and "down" refer to a characteristic called isospin carried by the quarks and given to the particle in which they reside. What exactly is meant by isospin is not important for this story, but in the case of the proton and the neutron it can take on two values, or orientations: up and down. Protons were said to have isospin up and neutrons isospin down. An extra up quark imbues the proton with that up isospin, while an extra down quark accounts for the isospin of the neutron. In Gell-Mann's model the presence of a third type of quark in each of the strange particles mentioned earlier caused them to behave differently from other particles. In light of that fact, he chose to call that third quark strange.

At the time, most physicists, including Gell-Mann, believed the quark model to be little more than an interesting scheme to categorize particles that explained some of the symmetries seen in the particle spectrum and might somehow lead to further insights. Of course, theoretical physicists are seldom too shy to think about things that don't yet have experimental evidence to support them. Before long, modifications were proposed to the quark model. Before the end of 1964, James Bjorken and Sheldon Glashow, two American physicists visiting the Niels Bohr Institute in Copenhagen, proposed the existence of a fourth quark. According to Glashow, "We called our construct the 'charmed quark,' for we were fascinated and pleased by the symmetry it brought to the subnuclear world."[16] At the time there were four known leptons: the electron, the muon, the electron neutrino, and the muon neutrino. Adding a fourth quark created a nice symmetry between the leptons and the quarks. In 1970, it was realized that the fourth quark provided a natural theoretical mechanism to avoid certain types of particle decays, which were expected in the three-quark model but were not seen experimentally. Theorists found the case for the fourth quark compelling on theoretical grounds—in spite of the fact that there was no direct evidence for the existence of even one quark, much less four different types.

Roughly a decade after quarks were first proposed, the first direct experimental evidence of quarks was uncovered in a series of so-called MIT-SLAC experiments at the Stanford Linear Accelerator Center (SLAC). In these experiments, an energetic beam of electrons was scattered off protons. The way in which the electrons scattered from the proton target provided information about what they hit. Imagine shooting bullets from a high-powered rifle at a car.

Bullets hitting the trunk and passenger regions of the car would pass through the car with relative ease. However, many of the bullets hitting the engine compartment would be scattered wildly as they ricochet off the engine block. The pattern of scatter for the bullets can provide information about the internal structure of the car, just as the pattern of scatter of high energy electrons can provide information about the internal structure to protons.

The MIT-SLAC experiments did not see quarks directly; rather, the way in which the electrons scattered off the protons showed that there was structure inside the protons that was consistent with being point-like, fractionally charged particles—just like quarks. The picture emerged slowly in the period from 1968 to 1972 as the experimental data was collected and analyzed. At the beginning of this period, the idea that quarks were real was out of fashion. It took a few years and some help interpreting the results from theorists like Bjorken and Feynman before the emerging picture became clear. Even then, there were skeptics.

The MIT-SLAC experiments involved a team of physicists led by two professors at MIT, Henry Kendall and Jerome Friedman, and a staff physicist at SLAC named Richard Taylor.[17] It was an exciting time, not without controversy. It was also a busy time for the physicists involved in the experiments. The room where the data was collected and monitored—the so-called "counting house"—became littered with rough plots of preliminary results generated quickly as the data rolled in. At one point in the midst of these experiments in 1970, according to Michael Riordan,

> It had become pretty obvious we'd have some exciting results. But these were rudimentary answers being generated by an on-line computer. Still to be applied were a number of important correction factors that could change the [results] substantially. Kendall began to get nervous that other physicists might come snooping around the counting house, glance at our rough plots, and make the wrong conclusions about what we were seeing. He therefore gathered them up into a

three-ring binder deliberately labeled "Administrative and Budget." Nobody, he figured, would ever want to peek inside that.[18]

On November 11, 1974, the dramatic discovery of a new particle by two independent groups was announced.[19] This particle was first seen in September 1974 in proton-beryllium collisions by an experiment led by MIT physicist Samuel Ting at Brookhaven National Laboratory. Ting named the new beast the *J* particle. Professor Ting and his team delayed announcing the discovery in order to make a series of careful checks to be certain of the find. In the meantime, a group of physicists from SLAC and Lawrence Berkeley Laboratory led by Burton Richter and Gerson Goldhaber independently discovered the particle in electron-positron collisions in early November 1974. They named the particle the "psi."

The new particle, called the "J-psi," as a testimony to the power of diplomacy—was very sharply defined in terms of its mass, which, according to the uncertainty principle, meant it was very long-lived as compared to the typical particles discovered up to that time. The fact that it was so long-lived and the dramatic, simultaneous discovery by two different groups led to an immediate acceptance worldwide that something new and puzzling had been found. The news spread like wildfire in the physics community. Physicists everywhere can tell you what they were doing when they heard the news. Riordan relates the story of James Bjorken hearing of the discovery from Burt Richter:

Early that afternoon [November 5, 1974] Bjorken was sitting down with his family for Sunday dinner when the phone rang. It was Richter with the startling news. "I couldn't believe such a crazy thing was so low in mass, was so narrow [in energy definition], and had such a high [probability of being produced]," he recalled. "It was sensational." He returned to the table a few minutes later, seemingly in a daze. He wife and children then watched open-mouthed as he unthinkingly heaped a large tablespoon of horseradish onto his baked potato and quietly began munching away, staring absent-mindedly off into space. "BJ," his wife finally counseled, "I think you'd better go down to the lab [SLAC] now."[20]

The announcement of the J-psi discovery heralded the beginning of roughly a two-year period physicists call the "November Revolution" in recognition of its importance. During this period, the particle physics community finally accepted the fact that quarks are real and that the odd characteristics of the J-psi and a number of similar particles discovered shortly after the J-psi announcement was due to the fact that they contained a fourth quark, the so-called charm quark predicted a decade earlier by Glashow and Bjorken.

This discovery of charm brought about more than the acceptance of quarks; it was also critical in the formation of the consensus behind the theoretical framework of the Standard Model. There were a number of experimental and theoretical developments in and around this time that provided evidence for the Standard Model. Among these developments were the results from the MIT-SLAC electron scattering experiments, the observation of so-called "weak neutral current" events at Fermilab and CERN,[21] and the resolution of the "R-crisis" through the quark model and a new theory of the strong interaction.[22] The addition of charm to these other bits of evidence was the thing that caused most of the skeptics to embrace the new paradigm. According to John Ellis, a leading theoretical physicist at CERN, "Charm was the lever that turned the world."[23]

The theoretical structure of the Standard Model evolved in parallel to the experimental developments of the late 1960s and early 1970s. It consists of two parts that are technically entwined but conceptually convenient to think of separately. One part is the theory of the strong interaction between quarks, which is known as the theory of quantum chromodynamics, or QCD for short. The other part is the so-called electroweak model, which incorporates the electromagnetic and weak nuclear forces in a single mathematical framework.

One of the most troubling aspects for the quark model in the early years was the fact that nobody had seen a quark. Even as the evidence for point-like structures inside the proton mounted, no bare, fractionally charged particles were seen directly in experiments. All of the evidence was indirect—as inferred, for example, from the way in which electrons scattered off protons. The reason for this shyness of quarks was eventually understood to be due to a fundamental characteristic of the strong nuclear force through which quarks interact with each other.

This force between quarks, as described by theory of QCD, is rather similar to the quantum theory of electromagnetism, QED. Both theories are quantum field theories, meaning the force comes from the exchange of a virtual particle. In the case of electromagnetism, that virtual particle is the photon; in the case of the strong force the particle that is exchanged is called the gluon. The electromagnetic force is felt by particles that carry electric charge. The strong force is only experienced by forces carrying *color* charge. There are two types of electric charge—positive and negative—and there are three types of color charge. The essential difference between the two theories is that whereas the photon is electrically uncharged, the gluon carries color charge. This turns out to create a very different type of force. In electromagnetism, the force between particles falls off quickly with distance. In the strong interaction, the strength of the force between two quarks increases dramatically as the distance between the quarks increases. This means that if one blasts two quarks apart, as they separate, the energy in the space between them grows to the point that quarks and antiquarks are formed from that energy and combine with the bare quarks moving apart such that the combinations no longer carry color charge.

This leads to the strange effect that if you blast apart a particle containing quarks, you see particles containing quarks coming out, but never a bare quark. On the other hand, when the quarks are in close proximity the strong force is rather weak. This means quarks can combine together happily—forming other particles like protons—so long as the total combination is color-neutral. The quark combinations that are found in nature are three-quark particles, known as baryons, and particles containing quarks and antiquarks, known as mesons, as shown in Figure 5.2.

The theory of QCD was developed in the early 1970s, just in time to play an important role in the November Revolution. As with most ideas, many people contributed. However, the critical understanding of how the force varies with distance was achieved in 1973 by a graduate student at Harvard University named David Politzer and a graduate student at Princeton University named Frank Wilczek, along with Wilczek's thesis advisor, David Gross. These gentlemen were awarded the 2004 Nobel Prize in Physics for this work.[24]

The other big part of the Standard Model is the electroweak theory, which is a quantum field theory developed in the late 1960s and early 1970s. A number of physicists contributed to this, including five who have been recognized through the Nobel Prize in Physics: Steven Weinberg, Abdus Salam, Sheldon Glashow, Martinus Veltman, and Gerardus t'Hooft.[25] As the name implies, this theory is a unification of the electromagnetic force and the weak force. The starting elements to this theory are independent, imperfect pieces— one that contains within it the basic characteristics of the electromagnetic force and a separate part that contains the basic structure of the weak interaction. At this stage the theory is far from perfect. Mathematical constraints in both the electromagnetic and weak parts of the theory require that all interactions take place between massless particles. To put it another way, the mathematical symmetry in the theory prevents the addition of terms representing mass in the theory. Because we live in a world of massive quarks and leptons, this is an unsatisfactory situation. The breakthrough comes when a new particle called the "Higgs" is added to the mix.

Has anyone ever seen or smelled or tasted a Higgs particle? No. Not yet anyway. But, it's fairly standard practice to hypothesize new things and see where they go in physics. Sometimes it leads to new and interesting insights. In this case, it led to very important results.

With the Higgs in the theory[26] the electromagnetic and the weak parts of the theory mix together mathematically, and massless quarks and leptons in the equations acquire terms that represent mass via interactions with the Higgs. In other words, adding the Higgs to the theory allows the quarks and leptons to have mass, leading some to say that the Higgs particle is the origin of mass. In addition, the theory predicts—in great detail—the existence and fundamental properties of three gauge bosons: the photon, which mediates the electromagnetic interaction, and the Z and W particles, which are exchanged in weak interactions between particles. The photon was known long before the invention of the Standard Model, of course. The W and Z bosons, however, are quite massive, and it was not until 1983 that these particles were finally produced in proton and anti-proton collisions at CERN.[27] The 1990s saw six collaborations[28] of particle physicists working at two new electron-positron colliders, sited at CERN and SLAC, respectively, study decays of the Z and W particles in excruciating detail only to find that the properties of these particles follow exactly the expectations of the Standard Model. Though the Standard Model was accepted broadly as an important step forward prior to this, the work at CERN and SLAC in the 1990s firmly established the Standard Model as the new paradigm.

The Higgs remains shrouded in mystery, though many physicists are working hard to change that as I write. Let's start with the name. The same group that brought you strange, charm, and bottom now give you the Higgs. What's with that? The Higgs particle is named after the English physicist Peter Higgs, who invented the theoretical technique that is now called the Higgs mechanism in 1964. That technique is a critical part of the Standard Model. Professor Higgs certainly deserves to have this hypothetical particle named after him. In fairness, though, there are other physicists who contributed significantly to the ideas surrounding the Higgs mechanism at around the same time. Recently, a Fermilab physicist named Ben Kilminster visited Rochester and gave a seminar in which he referred to the Higgs as the "BEGHHK" particle (pronounced as a guttural "beck") after the six main contributors to the ideas that became the Higgs: Robert Brout, Francois Englert, Gerald Guralnik, Carl Hagen, Peter Higgs, and Tom Kibble. It was done with a bit of humor because Hagen was in the audience. I doubt Kilminster's renaming of the Higgs will stick, but it was a nice gesture.

Okay. Stand up and stretch. Particle physics is always a bit of a tough go. There are too many names and particles to keep straight. Bear with me, though. The cosmological multiverse, if it exists, is filled with particles and forces like these.

Of course, the name is not the real puzzle about the Higgs particle. The real issue is that the Higgs has not been seen. Because this elusive beast is the key to the Standard Model and the Standard Model works incredibly well in describing three of the four forces of nature, it's a bit of a problem that we've not found direct evidence for the Higgs. Well, perhaps it's still too early to consider this a problem. After all, the Standard Model does not specify the mass of the Higgs, if it exists. Its mass could vary over a fairly wide range and still be suitable to make the theory work. Also, and perhaps more realistically, it is quite possible that there's no single Higgs particle as specified in the Standard Model. All we really know is that something in nature does what the Higgs does in the model. If, in fact, the Higgs were found to be exactly like it is hypothesized in the Standard Model, a great many particle physicists I know would be both disappointed and puzzled. A number of mysteries would not have a natural solution if that were the extent of things. For example, in the Standard Model quarks and leptons have mass, but their masses are not determined in the model. Why are the quark and lepton masses what they are observed to be? Why are there six quarks and six leptons?[29] The basic Standard Model does not answer these questions. There are other aspects of the Standard Model that require such a high degree of mathematical tuning that it leaves physicists feeling uncomfortable.

Many physicists consider these troublesome issues with the Standard Model more as opportunities than problems. They indicate our picture is incomplete and that there are many good reasons to expect new physics—physics beyond the Standard Model—that will supply something that plays the role of the Higgs and, hopefully, solve some of these other important questions left unanswered by the Standard Model.

This situation is nothing new, really. Physicists have been seeking physics beyond the Standard Model for the last 40 years. Amazingly, the Standard Model has withstood this onslaught with incredible resilience, as most of these experiments have done little more than probe and confirm the Standard Model in pain-staking detail.

Though the fundamental aspects of the Standard Model have withstood the experimental testing to date, it has been necessary to make some modifications in the Standard Model to accommodate new discoveries. For example, within the Standard Model there is a natural way to group the particles into a unit known as a "generation." One generation consists of the electron, electron neutrino, up quark, and down quark. The second generation is made up of the muon, muon neutrino, charm quark, and strange quark. The basic paradigm of the Standard Model evolved with these two generations. However, a third generation was discovered and added to the model. This third generation consists of the tau lepton (discovered at SLAC in the period between 1974 and 1977[30]) the tau neutrino (discovered at CERN in 2000), the bottom quark (discovered at Fermilab in 1977), and the top quark (discovered at Fermilab in 1995). This third generation of particles fits quite nicely into the Standard Model.

A second area where the Standard Model has evolved since 1974 involves the differences between matter and antimatter. In the basic Standard Model, it was assumed that particles and antiparticles have very similar, but mirrored properties—that is to say, for example, a particle and its antiparticle decay with exactly the same probability. This turns out not to be the case all the time. There are tiny differences between how matter and antimatter behave. Interestingly, the Standard Model can accommodate such differences provided there is a third generation of particles—as we know is the case.

Yet a third place where the Standard Model has been modified slightly is that neutrinos, once thought to be massless, are now known to have a small mass. Again, this is something that has been fit into the Standard Model structure with relative ease.

The discoveries mentioned that came about since the end of the November Revolution have been accommodated with small changes in the Standard Model and no major revision in our thinking. Nevertheless, we know the Standard Model is incomplete and there is hope that further probing will provide clues as to how to move beyond the Standard Model. The Large Hadron Collider (LHC) program at CERN was initiated, at least in part, to look for evidence for the Higgs or what does what the Higgs does. Other programs are making more and better measurements of the matter-antimatter differences and neutrino properties. Hopefully, a surprise awaits us around the corner.

Interestingly, the most compelling evidence we have currently for physics beyond the Standard Model comes not from what we see, but rather from what we do *not* see. Oddly enough, the first indication of this—what we now consider a new and exciting possibility—came in 1933, some 35 years before the creation of the Standard Model. A rough-edged astrophysicist at Caltech named Fritz Zwicky made a careful study of a cluster of galaxies in the constellation Coma Berenices. In particular, he compared the mass of the cluster determined from the brightness and distance of the galaxies to the mass determined from something called the virial theorem. The virial theorem relates the average speed of the galaxies to the extent to which each galaxy is attracted to the others by gravity, which depends on the overall mass of the cluster. Both techniques used by Zwicky to measure the mass of the galactic cluster were relatively conventional and used with confidence in other instances. Yet, Zwicky discovered a striking disagreement between the masses derived using the two different techniques: He found the mass determined from the luminosity of the galaxies to be a factor of 10 too little to account for the motion of the galaxies. Zwicky called this missing mass "dark matter" in recognition that it was unassociated with the mass generating the visible light coming from the cluster of galaxies.

Fritz Zwicky was a very creative scientist. In addition to the discovery of dark matter, Zwicky was the first person to hypothesize the existence of neutron stars. He surmised, correctly, that instabilities in the latter stage of a star's life might lead to an enormous explosion leaving behind a core of nuclear matter—essentially neutrons. Zwicky and his collaborator, Walter Baade, coined the term "supernova" to describe this massive stellar explosion.[31]

Zwicky also had one of the crustier personalities in the history of science. He is famous for coining another phrase. He described many of his colleagues as "spherical bastards," meaning they were bastards no matter how you looked at them.[32] It seems that Zwicky's usage of this and similar terms was so common that once when some of his graduate students arrived at his house for dinner one evening, it is said Zwicky's wife answered the door and yelled back into the house, "Fritz, the bastards are here!"[33]

For 40 years Zwicky's result on dark matter stood as a puzzle. Then, in 1975, two astronomers from the Carnegie Institute, Vera Rubin and Kent Ford, presented a result at an American Astronomical Society meeting that caused a stir. In this work they showed that the rotation speed of stars about the cores of spiral galaxies implied that spiral galaxies contain substantially more mass than was expected from the visible mass seen in stars and luminous gas. Again, the discrepancy was not small; it was a factor of 10. The results of Rubin and Ford were initially greeted with skepticism, but the quality of the work and confirmation by others soon led to a general acceptance of the work and the recognition that the majority of the mass locked up in galaxies is not visible except through gravitational effects.

The evidence for the existence of dark matter has grown through the years.[34] It seems that we do not understand the fundamental nature of approximately 80 percent of the matter in the universe. This is generally recognized by physicists as one of the most intriguing and important scientific puzzles of our day.

To say that we don't understand the nature of dark matter is not to say that we know nothing about it. Black holes and stars called brown dwarfs, which never acquired enough mass to initiate fusion reactions, are both sources of mass that do not contribute much in the way of light to their surroundings. Neutrinos are also a form of dark matter. They exist in abundance, have a small mass, and cannot be seen except through weak interactions. Still, black holes, brown dwarfs, and neutrinos are thought to provide only a small fraction of the mass necessary to solve the missing mass problem. The bulk of the dark matter, which is the majority of the matter in the universe, has certain characteristics inferred from its interaction or, rather, lack of interaction, with other things. Dark matter is expected to be electrically neutral. Otherwise it would interact with light. A physicist would say dark matter does not interact electromagnetically. Dark matter also appears to not feel the strong nuclear force and cannot be made up of protons and neutrons. In other words, it does not appear to be normal matter in some hidden form. If this were not the case, the abundances of the elements formed in the early stage of the evolution of our universe (discussed more later in the book) would be different from that which is observed. There are indirect indications that dark matter—whatever it is—may interact very weakly with normal matter through some force, in addition to gravity, that is not yet understood. This weak interaction provides a convenient way for theories of the early universe to tune the amount of dark matter we see in the universe left over from the early Big Bang. This reasoning leads many physicists to talk about the hypothetical dark matter particle as a Weakly Interacting Massive Particle, or WIMP.

I kid you not. Dark matter particles may be WIMPs. I have very talented and brilliant friends working hard on experiments designed to detect WIMPs. This is sure to lead to amusing headlines if they succeed in discovering them!

One exciting possibility is that there's a connection between the Higgs problem and dark matter. One theoretical idea that is compelling to many physicists and, potentially, might solve both the Higgs riddle and provide a dark matter candidate is called "supersymmetry." According to theories exhibiting supersymmetry there is a deep symmetry in nature centered around a characteristic of particles called "spin."

The concept of spin was invented to account for the fact that many particles have small magnetic fields much like a tiny bar magnet. The use of the word *spin* comes from the fact that a spinning ball of electric charge also creates a magnetic field that resembles that of a little bar magnet. It turns out that particle spin is a bit more complex than that naïve mechanical model would imply. Neutrons are neutral particles and they have spin, for example. Rather than delve into the full complexity of spin, let's consider only a few facts about spin that are relevant for our story: Each particular type of particle (electrons, quarks, photons, and so on) has a characteristic spin; spin is quantized, meaning particles can have only particular values of spin; in the correct unit of measurement, particles can have spin with a value of 0, $\frac{1}{2}$, 1, 3/2, 2, and so forth; particles with integral spin (0, 1, 2, 3, and so on) are called "bosons"; and particles with half-integral spin ($\frac{1}{2}$, 3/2, 5/2, and so on) are called "fermions." Examples of bosons are the photon, Z, and W particles. Examples of fermions are the electron, muon, neutrinos, and quarks. This categorization is important theoretically, because fermions and bosons have very different quantum behavior. In particular, one useful difference is that quantum loop corrections for fermions and bosons enter calculations with opposite signs.

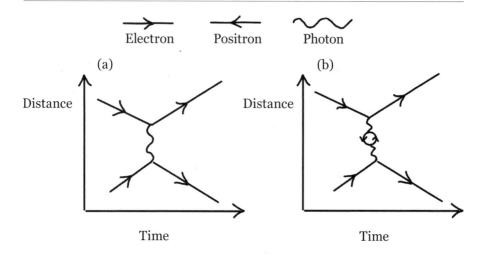

Figure 5.3: Feynman diagrams illustrating quantum loop corrections to a particle interaction. Figure (a) illustrates a basic electromagnetic interaction between two electrons through the exchange of a virtual photon. In (b), another quantum possibility is shown. Here the virtual photon decays into an electron-positron pair, which then turns back into a photon. The loop in the virtual photon is called a loop correction and other particle-antiparticle pairs can contribute similar corrections that must be taken into account.

We briefly encountered quantum loop corrections earlier, but I didn't call them by that name. Recall that the quantum picture of a particle or a process is complicated by the fact that the uncertainty principle allows virtual particles to be present briefly during a very short snapshot of time. Any rigorous calculation has to take into account all of these virtual, quantum possibilities. For example, consider the Feynman diagrams in Figure 5.3.

Figure 5.3(a) shows the basic Feynman diagram for an electron being repelled by—or scattering from—another electron. The two electrons are exchanging a photon, which is the gauge boson that mediates the electromagnetic force. Of course other things can happen so long as the uncertainty principle is satisfied. For example, the process shown in Figure 5.3 (b) is also a possibility, where the virtual photon changes momentarily into an electron and positron that quickly annihilate back to a photon. In this diagram, the loop in the middle of the photon being exchanged is called a quantum loop and the way in which this affects the calculation in the end is called a quantum loop correction. There are many other possible diagrams that could happen. Other loops might be present or some other fermion might be the particle in the loop. For example, instead of an electron and a positron in the loop, it could be a muon and an anti-muon, or a quark and an antiquark, and so forth. A theorist wanting to calculate the nature of the scattering of one electron from another would need to take into account all of these quantum possibilities. Fortunately, not all the possible diagrams contribute equally to the calculation. Most of them can be safely neglected because their contribution is very small. One of the things a theoretical physicist learns in graduate school is how and when certain diagrams can be neglected in a computation.

One of the troublesome aspects of the Standard Model is that quantum loop corrections that come into calculations of the Higgs mass naturally tend to make the Higgs mass very large—too large, in fact, for the Standard Model to work the way it should to agree with nature. Because of this it is necessary to fine-tune the quantum loop contributions to the Higgs mass in an unnatural way in order for the Standard Model to work. This is known as the "hierarchy problem."

The theoretical issues surrounding the hierarchy problem can be improved by hypothesizing a deep symmetry between fermions and bosons. The idea is to assume that for every type of fermion (boson) in nature there is also a boson (fermion) that is similar in every way except for the spin. If this were true, then quantum loop contributions to the Higgs mass involving a fermion (boson) would be cancelled by those involving its boson (fermion) partner, because fermions and bosons contribute to the quantum loop corrections with opposite signs. Theories including this fermion-boson symmetry are called supersymmetric theories.

We know that nature is not truly supersymmetric. If it were supersymmetric, not only would we see normal electrons, which are fermions, but we would also observe supersymmetric electrons—particles that look like electrons in every way except they would be bosons because they would have integral spin. In a truly supersymmetric world, all the other normal particles would have supersymmetric partners, too, leading to a doubling of the known particle spectrum.

This doubling of the particle spectrum is not seen, of course. That's okay. Intrepid theorists have found a way around that little difficulty with supersymmetry. The idea is that nature is basically supersymmetric at high energies but that as low energy the symmetry is imperfect. They say that something *breaks the symmetry*, leaving the supersymmetric partners to the normal particles having very high masses. We don't see them because all the processes we have observed to date have too little energy to produce such massive particles. In this case, if supersymmetry were true, we should expect to discover an entirely new spectrum of particles as some high energy scale—perhaps in collisions at the Large Hadron Collider.

If nature is supersymmetric the hierarchy problem with the Higgs can be solved.[35] In addition, extensions of the Standard Model with supersymmetry have no problem supplying a Higgs—or even more than one Higgs—to do what the Higgs does in the Standard Model.

Finally, the connection with dark matter comes in because many popular supersymmetric theories have a lightest supersymmetric particle that cannot decay into normal matter. Because there isn't anything else for it to decay into, it is absolutely stable. If it is also electrically neutral, this lightest supersymmetric particle is an ideal candidate to be the particle that makes up the bulk of the dark matter in our universe.

The Standard Model has proven to be an incredibly successful scientific theory that has shown we have great insight about nature.

Yet there are many remaining questions that come to mind: What is dark matter? What drives the Higgs mechanism in the Standard Model? Is it a Higgs or something new? What determines the spectrum of the leptons and the quarks? Is nature supersymmetric? And on and on.

Physicists are working hard to discover answers to these questions. Perhaps the most visible effort that may shed light on these issues is the Large Hadron Collider project at CERN. This accelerator is colliding protons with protons at energies not yet studied extensively in the laboratory. The LHC collisions might have enough energy to make the Higgs and/or supersymmetric particles and/or dark matter particles and/or a complete surprise. In addition to the LHC there are many smaller efforts to push beyond the Standard Model (too numerous to mention here).

It's a grand adventure, and I've little doubt that nature has some interesting surprises in store for us. If I were to bet on it though, I'd probably put my money on supersymmetry. Why? Well, if supersymmetry were discovered to be true, we'd be stuck with a whole series of additional silly particle names. The superpartners of fermions are given the fermion name with a "s" in front. So, the supersymmetric partner to the electron is called the selectron and the partners to the muon, neutrino, and quark are the smuon, sneutrino, and squark, respectively. My favorite of these is the top squark because it is often referred to simply as the stop. And then there are the supersymmetric partners of the bosons. These particles are distinguished from the normal particles by the addition of "ino" to the end of the normal particle name. So the supersymmetric photon is the photino and the supersymmetric partners of the Z and the W are the Zino and the Wino, respectively. How can this theory possibly be wrong if it leads to a Nobel Prize for the discovery of the WIMP that is a Wino?!

Finally! With relativity, quantum mechanics and particle physics under our belt, we have in place the raw materials to discuss cosmology and the many fascinating cosmology-derived multiverse concepts. When I was in graduate school, cosmology seemed sort of like a far-fetched business. The experimental data were very poor, so it was hard to imagine cosmological science progressing very much. That's really no longer the case. Cosmology is now a precision science with a trophy case full of very exciting recent discoveries. So, hang on for a fun ride.

6

A Case of Cosmic Acne

A couple of years ago I received an e-mail asking me if I was "the Steve Manly"—the colonel in the Space Rangers. I was greatly concerned. How had I been found out?! As it turns out, a couple in Oregon was fixing up a baby nursery in 1950s space retro and stumbled across a series of posters entitled "The Astonishing Adventures of Colonel Steve Manly Space Ranger." They did a Web search on the name to try to learn more about the posters. They contacted me thinking that perhaps one of my old physics students pulled an entrepreneurial joke with the poster. I was thrilled to find out about the posters and eventually traced them to their source. I discovered that the name was more or less pulled out of thin air during the poster design.[1]

I might have been even more excited by this if I'd grown up in the 1950s, but the whole Space Ranger thing was before my time. I grew up when life in space was rapidly becoming a reality. I remember the incredible excitement of the *Apollo* missions. Like most kids of that time those amazing moon missions ignited my imagination and my interest in outer space. My dad prodded those interests a bit when gave me a small telescope while I was in middle school. More than the telescope, though, it was the accompanying subscription to *Astronomy* magazine that really piqued my interest in the strange things far beyond our Earthly shores. The beautiful photographs and articles about strange things like black holes inspired me. Those beautiful photos also fostered in me quite an inferiority complex about my little telescope. You see, it wasn't much of a telescope, really. It had a 4-inch diameter mirror and barely allowed me to see Saturn's rings on a good night. Things seen with my little scope didn't look much like those amazing images in the magazine.

As with many inferiority complexes, this one had staying power. Some 25 years later, I decided that it would be a good educational experience for my own children if I bought a telescope. That's the way it works, right? Use the kids as an excuse to buy the toys you want? At least it wasn't a boat. After looking around a bit, I was quite impressed with how much scope you could get for the money. So, I determined just how big I could go and still physically manage to cart the telescope out into the yard on a clear night. I bought a telescope with a 12½-inch diameter mirror. The thing is about the size of a decent household water heater (a fact of which I am reminded by my wife every time spring cleaning rolls around).

My toy did wonders for my reputation around the neighborhood. It arrived—predictably—during Rochester's winter when clear skies are seldom seen. So, for the first month, my telescope lay on the living room floor while I tinkered with aligning the optics periodically. During this period my wife and I hired one of the neighborhood high school kids as a babysitter and went out for the evening. Months later I heard that the sitter returned home that night telling her mother, "I don't know what it is, Mom. But it looks like a missile!"

I chose to study inner space—particle physics—for a career in spite of my love affair with outer space. At the time I thought I preferred to study something in the lab, where it was possible to do controlled experiments. There is no control over things in space, which makes it difficult to do science. Though it may have played a role in my professional direction, this reasoning is only partially correct. The inability to run controlled experiments does hamper the scientific process for those seeking to understand astrophysical and cosmological phenomena. Still, with powerful instruments, patience, ingenuity, imagination, and the huge number of examples of each type of object to study, astronomers and astrophysicists have made incredible scientific progress in the past 100 years. They've developed a robust scientific picture of our universe and the objects within it. To me, perhaps the most exciting aspect of this work is the realization that there's an intimate connection between the world of inner space, and the structure and function of things on astrophysical and cosmological scales. It turns out that I wasn't making the choice I thought I was years ago. The very large and the very small are inextricably entwined, and the study of either involves both. This connection between the small and the large plays a critical role in most of the cosmological visions of the multiverse.

Arguably, modern cosmology began around 1912 when the American astronomer Vesto Slipher measured the relative velocity of a handful of galaxies with respect to our galaxy. He calculated these velocities by determining the so-called "redshift" in the optical spectra of each of the galaxies. This technique, based on the Doppler effect, has proven to be greatly important in astronomy and cosmology through the years. The Doppler effect is the change in the frequency of a light wave that comes about when there is relative motion between the light emitter and observer. The effect on light is too small to be noticed in typical daily activities. There is a similar effect on sound, as it is also a wave phenomenon. The effect on sound is quite noticeable in daily life. In fact, it is so common as to often pass unnoticed. If you stand by a highway with fast-moving cars, the pitch of sound produced by approaching cars is higher than the pitch of the sound produced by the same cars as they recede. In the case of light, the color changes. Light emitted by an approaching object is bluer—shifted to a higher frequency—than it would be if the light were emitted by an object stationary with respect to you. In the lingo of an astronomer, the light is blueshifted. For a receding object the light is shifted to lower frequency, or redshifted. The extent of the blueshift or redshift depends on the speed of approach or recession of the object, respectively.

The operation of the radar gun used by policemen in speed traps is based on the Doppler effect. In this case, the gun sends out a radio or infrared light beam and measures the change in the frequency of that beam as it reflects off the target vehicle. The degree to which the reflected frequency differs from the original frequency can be converted to a reliable speed for the vehicle in question. If you plan to fight a speeding ticket, don't bother questioning the Doppler effect. It's quite well understood and can be used effectively for this purpose.

The speed trap technique won't work for astronomical objects, however, because it's not very practical to bounce beams off distant stars and galaxies. Instead, astronomers rely on a built-in frequency scale supplied by nature. Recall that each atomic element has a unique set of energies available for the electrons to occupy. Consequently, the pattern of discrete frequencies of light emitted and absorbed by each type of atom is unique. Those patterns of so-called spectral lines are known to scientists and are used to identify the presence of particular types of atoms. Slipher observed the spectral lines emitted by atoms in the distant objects and compared the observed frequencies to those emitted by atoms at rest with respect to him. From the extent to which the spectral lines were shifted in frequency, he determined the velocity of the object with respect to Earth along the line of sight. Modern astronomers continue to use this technique.

Vesto Slipher measured the frequency shift in the atomic spectra of light coming from a number of nearby galaxies. Having done so, he noticed something strange: The vast majority of the galaxies were receding from us. This is not something you would expect to find if the universe were fundamentally static and populated with galaxies moving in random directions.

In the early 1920s, an astronomer named Edwin Hubble used what was then the world's most powerful telescope to determine that our universe is made up of vast islands of stars—galaxies—separated by huge distances. Astronomers had observed many other galaxies before Hubble's work, but there was an ongoing dispute as to whether or not these objects resided inside or outside our own galaxy. In one school of thought, the Milky Way was the extent of our universe. Using the powerful Hooker Telescope at Mount Wilson Observatory in California, Hubble was able to see individual stars in nearby galaxies. In particular, he observed examples of a class of star known as a Cepheid variable that oscillates in brightness with a period that depends on the absolute amount of light emitted by the star. By measuring the timing of the variation in brightness, Hubble was able to infer the absolute brightness of the star and, from that, he could tell the distance to the star by the apparent or observed brightness of the star—just as you could determine the distance to a candle by observing how bright it appears on a dark, clear night.

In 1929, Edwin Hubble and a talented observatory-janitor-turned-staff member named Milton Humason combined Slipher's redshift measurements with Hubble's measurements of the distance to the different galaxies and realized that the redshift to distant galaxies increases in proportion to the distance to the galaxy. This observation—that galaxies further from us are moving away from us faster in direct proportion to their distance—has come to be called Hubble's law.

(a)   (b)

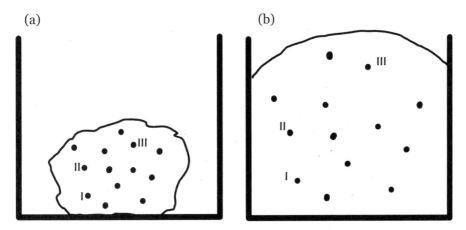

Figure 6.1: Raisin bread dough before (a) and after (b) rising. Note that the distance between raisins I and II has increased less than that between I and III in the same amount of time. This is because in the expanding dough, as in our universe, the recession speed is proportional to distance.

Hubble's law is the relation you expect to have if the space in the universe is expanding. Figure 6.1 illustrates this. Imagine a ball of raisin bread dough sitting on the counter before it rises as shown in figure 6.1(a). Note the distances between one raisin (I) and two others, one close (II) and one far away (III). Figure 6.1(b) shows the same bread after it rises. Because the bread is expanding throughout, each raisin recedes from all the others. Again the distances to the same two raisins are shown and, of course, that distance has increased in both cases as the bread rose. But note that because more of the expanding dough was between I and III, the distance between I and III is now much larger than the distance between I and the closer raisin, II. Because raisin III moves farther from raisin I than does raisin II in the same time, the recession velocity of raisin III with respect to raisin I is greater than that of raisin II. In other words, during the expansion, the raisins all moved apart, and the recession velocity of one raisin with respect to another raisin was proportional to the distance between them—Hubble's law for a raisin bread universe.

As strange as the concept of an expanding universe may seem, Hubble and Slipher were not the first people to propose the idea. A Russian physicist named Alexander Friedmann discovered the expanding universe solution to Einstein's equations of general relativity in 1922. Einstein himself, in what he would later call "the biggest blunder of my life,"[2] inserted a constant in his equations in 1917 in order to keep the universe static. The static universe was, to Einstein, a constraint in the problem. There was no evidence to support an expanding universe at the time and he felt a non-static universe was an unacceptable solution. In 1927, a Belgian Catholic priest and physicist named Georges Lemaître published a paper entitled "A homogeneous Universe of constant mass and growing radius accounting for the radial velocity of extragalactic nebulae," in which he presented an expanding universe solution to general relativity similar to that which Friedmann found five years earlier. At the time, the word *nebulae* was used to describe nebulous things in the sky ranging from galaxies to clouds of gas. In this title, it refers to galaxies.

Interestingly, Lemaître was aware of the work of Slipher and others showing that redshift measurements indicated that most galaxies are receding from the Milky Way. In fact, in his 1927 paper, Lemaître tied together the redshift measurements with his view of the expanding universe and derived the relationship between redshift and distance that was to later become Hubble's law. Unfortunately for Lemaître, he published in a Belgian journal that was not widely read and most physicists, including Hubble, remained ignorant of the work until 1930.

Lemaître reasoned that if the universe is expanding and one works backward in time, the universe must once have been very small and have originated at some finite time in the past. In 1931, Lemaître proposed that the universe originated from a "primeval atom" that exploded, giving rise to space and time and the expanding universe. This is the idea that was christened the "big bang" by the English astronomer Fred Hoyle in a 1949 radio show.

The name "big bang" is something of a misnomer. The space itself is expanding. There is no explosion at the beginning of the expansion, because there is no space for anything to explode into. When I get to this topic in class and my students look sleepy I tell them that the big bang is not so much like a popping zit as it is a butt that grows from within as you add weight. It's a bit graphic, but it gets the point across and awakens them.

The expanding universe solution to general relativity that is the basis for the big bang model is now called something like the FLRW model for Friedmann-Lemaître-Robertson-Walker or the Roberson-Walker metric or the Friedmann-Lemaître model, depending on what is being discussed or who is doing the talking. Friedmann and Lemaître we have discussed. Howard Robertson and Arthur Walker are a couple of other scientists who contributed to our understanding of this theory in the 1930s.

Whatever it is called, this solution to general relativity does not, by itself, constitute what we mean by the big bang model. The FLRW model is a solution to the equations of general relativity for a homogeneous and isotropic (the same in every direction) expanding universe. Other things need to be incorporated into the picture to get a realistic model of the evolving universe. For example, homogeneous and isotropic sounds great, but our universe has big lumps in it called galaxies and planets and dogs. From whence did all that structure come? Even if you argue the galaxies are tiny on the scale of the universe, FLRW says nothing about how exactly the smoothly distributed matter came to be and whether it's made of quarks or cotton candy. Even if you knew about quarks and leptons, things are much more complex than tossing the parts into a bucket and mixing them up. The universe is cooling off as it expands in much the same way gas expanding out of a spray nozzle cools. Extrapolating back in time, this means that the universe was once vastly hotter than it is today. Temperature is a measure of how fast the particles in the gas are moving and the degree of violence in the collisions between them when they happen. At higher temperatures, what is stable in a cool environment can no longer hold together, large masses become irrelevant in the collisions, and the fundamental forces between particles change. All of this must be part of the model for the theory to be taken as a serious cosmology.

Many physicists played important roles in fleshing out the details of what we now call the hot big bang model, and I will mention only a few of the names as we move along. The hot big bang theory is the core of the standard model of cosmology today. Most of it is well-accepted in the physics community because it is supported by the data.

How can the entire universe have started at a point? What banged? Can physics as we know it describe the infinite energy density that must have existed at that point? These are all good questions. There's no consensus about the earliest fraction of a second where the direct scientific evidence is sparse and the physics we understand well breaks down. Experimental probing of our universe continues, as it is science in progress. So, perhaps even this earliest stage will be well understood someday. For now, don't get hung up on what happened at the very first instant of creation in the big bang model. There's much to be learned from the moments just after and beyond. We'll come back to the very start—the bang—in the next chapter because a number of exciting multiverse scenarios are tied up in exactly what and how things happened in that first mysterious fraction of a second. For now, though, let's concentrate on the parts of the story that are on a fairly solid footing scientifically.

To understand where the universe is today and where it will be in the future, we must study the past. The universe becomes hotter and denser as we push back in time, and to understand how the universe evolved we must know what was in the universe at that time and understand the forces of nature under those conditions.

Frankly, at first glance, this task of knowing the past seems impractical, if not impossible. After all, we'd need a time machine in order to go back and study the past. Happily, it turns out that the past is not as inaccessible as it might seem. We have *two* different time machines at our disposal. Our first time machine is the telescope. The finite speed of light means that it takes time for light from distant objects to reach Earth. As we look farther out in space, we are looking at processes that took place in the universe farther back in time. Our other time machine is the high energy nuclear and particle accelerator, which allows us to study the nature of matter and forces at energies that were prevalent in the early stages of the big bang and in astrophysical processes like supernovae. In addition, we have the *present*. Just like a middle-aged hockey player carrying the scars of his youth, the universe as we know it is a product of its past history. So, physicists use knowledge about the current universe to constrain what they think could have happened in the past.

We can extrapolate our knowledge further back in the past than you might imagine. Currently, it is believed that the big bang occurred roughly 13.7 billion years ago. The energy achieved recently[3] in collisions at the Large Hadron Collider (LHC) correspond to that expected to be present typically in the universe at roughly 10^{-14} seconds after the start of the big bang. Eventually the LHC will probe energies corresponding to times almost a factor of ten earlier than that—to 10^{-15} seconds after the big bang. So, experiments here on Earth enable us to probe physics that is relevant to virtually the entire history of the universe. There are, of course, some things in there that we don't understand yet, such as dark matter and the Higgs mechanism and at least one more surprise that I've not discussed yet that physicists call dark energy. Still, there is much that we *do* understand that is applicable throughout the history of the universe except for the tiniest fraction of a second right after the big bang. To be fair, though, that tiny fraction of a second is enormously important and determines largely what happens during the rest of the time, but we'll ignore it for now.

The hot big bang model does not claim to solve the problems of what happens during the first very tiny fraction of the first second after the start of the big bang, or what came before that. It does, however, provide a picture of the universe from about 10^{-15} seconds on, where the conditions are directly accessible for study using accelerators here on Earth.

At 10^{-15} seconds after the start of the big bang, or $t=10^{-15}$, the universe was an extremely hot, smooth, primordial soup populated by elementary particles. The typical energy for particles in the soup was about 7 TeV. In case the unit I'm using seems odd to you, an "eV" stands for electron-Volt. One eV is the amount of energy gained by an electron as it moves from one terminal to the other terminal of a 1-Volt battery. The mass of an electron is 0.511 million electron-Volts, or 0.511 MeV. One TeV stands for one Tera electron-Volt, or one million MeV, or one million-million electron-Volts.

Okay, so maybe all these gazillions and electron-Volts are a bit much. We could try a different measure of the energy. The temperature at 10^{-15} seconds after the start of the big bang is thought to have been around 10^{17} degrees. For an amusing comparison, the core temperature of the sun is approximately 10^7 degrees, or 10 billion times cooler than the universe as a whole was at that $t=10^{-15}$ seconds.

Of course, there was no sun or any other star around at 10^{-15} seconds after the big bang. It was much too hot for nuclei and atoms to exist. The temperature was so high that the typical collision energy was much higher than the particle masses. This means heavy particles were produced as often as light particles. The universe was a hot gas or soup of quarks, gluons, leptons, photons, and W and Z particles. At such an energy density the strong force is too weak to bind quarks into hadrons. So, there were no protons and neutrons or pions at that point.[4]

As time passed the universe expanded and the temperature dropped. By t=10^{-4} seconds, the temperature dropped to around 10^{12} degrees and the typical interaction energy was 100 MeV. At this energy, the strong nuclear interaction becomes strong enough that it binds the quarks and gluons into protons and neutron and other mesons or baryons. This very brief period in the history of the universe is called "baryogenesis" because it is when the protons and neutrons, both of which are baryons, were first made.

During the first few minutes after the big bang, the universe continued to expand and cool—down to a frigid billion degrees. In these few minutes some of the protons and neutrons began to stick together, and a few light nuclei like deuterium and helium and lithium were formed. This process is known as "big bang nucleosynthesis."

Think of the universe at this point as a big cooking pot full of protons and neutrons. The density was fairly high and as the pot cooled the protons and neutrons began to stick together forming deuterium nuclei. Initially the deuterium nuclei—consisting of a single proton and a single neutron—were blasted apart about as quickly as they formed because collisions were still too energetic for that relatively fragile nucleus to hold together. Around 100 seconds after the big bang, the universe cooled enough for the deuterium to hold together. This is important because the deuterium acted as a building block for creating heavier elements like helium, which consists of two protons and two neutrons. There was only a short window of time during which the deuterium was stable and the collisions had enough energy to power the cascade of fusion necessary to create helium and a few other light elements. At the end of three minutes, the universe cooled to the point that fusion reactions stopped and the abundances of the light elements in the universe were fixed. Most of the normal matter in the universe at this time (we aren't mentioning dark matter in the context of the big bang yet) was in the form of free protons, and around a quarter of the matter, by mass, was in the form of the nucleus helium-4 (two neutrons and two protons). Then there was some deuterium and tritium and lithium, all in smaller amounts.[5]

The theory for how the elements in the universe were formed in the early big bang cooking pot was worked out in the late 1940s and early 1950s. The initial work on this was done by Ralph Alpher, then a graduate student, and George Gamow, who was his research advisor at Johns Hopkins University. It was the subject of Alpher's PhD thesis. When it came time to publish the work, Gamow added the name of a friend at Cornell University, Hans Bethe, to the author list in spite of the fact that Bethe had nothing at all to do with the work. Bethe, of course, knew nothing of the paper until he saw it in print, but he appreciated Gamow's joke. Instead of Alpher and Gamow, the paper was authored by Alpher, Bethe, and Gamow, which was a delightful play on the Greek letters alpha, beta, and gamma. The paper is known as the alphabetical paper or the alpha, beta, gamma paper.

This period of nucleosynthesis and the theory behind it plays an important role in the story because it provides us with a compelling bit of direct scientific evidence for the big bang. All the nuclear reactions that drive it have been studied in the laboratory. The theoretical predictions for the elemental abundances in the early universe—the so-called primordial abundances—are robust and physicists are confident in the calculations. Making measurements to test these predictions is difficult because fusion reactions in stars have contaminated the primordial element distribution by throwing other nuclei, not from the time of the big bang, into the mix. But, astronomers and cosmologists have found ways to make these measurements, and the hot big bang nucleosynthesis theory is in good agreement with the data.

The early universe was opaque. There was plenty of light around; that wasn't the problem. Photons were bouncing around all over the place. The issue was that the temperature of the universe was so hot that neutral atoms could not form. The instant an electron and proton joined to form a hydrogen atom, the electron was knocked free. Consequently, the universe was a gas of charged particles—a plasma—and in such a gas, photons can't travel very far before being scattered. So, even though the universe was very bright, it was opaque, a bit like a frosted window. This opacity is important to us because no matter how powerful a telescope we may build, we will never see a clear image from this opaque period of the universe.

However, by the time the universe was several hundred thousand years old, it had cooled to a temperature of around 3,000 degrees. At this stage, all the charged nuclei and electrons combined into neutral atoms, mostly hydrogen, and the universe became transparent. Most of the photons in the universe at the time when this happened have traveled freely through the transparent medium of our universe since that time. Physicists call this particular transition in the universe "recombination," though the name seems a bit misleading because the electrons and protons were not combined before that. The word comes from plasma physics where physicists study plasmas in the laboratory.[6]

Of course, not all the photons emitted at the end of recombination will travel forever. A few of them have hit Earth. This light traveling to us from the time of recombination some 13.3 billion years ago is called the Cosmic Microwave Background (CMB) radiation. Looking out from Earth in all directions at this light, scientists have created a map of the sky that provides us with a detailed image of the universe at that time—some 380,000 years after the big bang. There is a great deal of information about the early universe hidden in this image. In fact, the CMB has been called the cosmic Rosetta Stone because of the clues it holds that help us decipher many aspects of the evolution of our universe.[7]

The existence of the CMB was predicted in 1948 by Gamow and Alpher, along with a collaborator named Robert Herman, and then predicted again in the early 1960s independently by a Russian, Yakov Zel'dovich, and two Americans at Princeton, Robert Dicke and Jim Peebles. The idea that something like this can be predicted or discovered independently multiple times may seem a bit strange. But, scientists are people and literature searches were a bit more troublesome in the days before on-line search engines. In the words of Dicke:

> There is one unfortunate and embarrassing aspect of our work on the fire-ball radiation [CMB]. We failed to make an adequate literature search and missed the more important papers of Gamow, Alpher and Herman. I must take the major blame for this, for the others in our group were too young to know these old papers. In ancient times I had heard Gamow talk at Princeton but I had remembered his model universe as cold and initially filled only with neutrons.[8]

According to the theoretical expectations, the CMB has very particular identifying characteristics. For example, the universe at the time of recombination was thought to be a uniformly distributed hot gas at a temperature of around 3,000 degrees. The spectrum of light at that time should have looked exactly like the glow of light from an object at that temperature. In other words, it should look like Planck's blackbody spectrum emitted by an object at that temperature, which happens to have a peak frequency in the infrared part of the electromagnetic spectrum. But, there is a complication that must be taken into account: The CMB photons we observe today have been traveling for some 13.3 billion years. From our perspective today, they were emitted from a bit of the universe that is very distant from us. By Hubble's law, we perceive light emitted from such a distant object to be greatly redshifted. In fact, this leads to the expectation that the peak of the CMB radiation will fall in the microwave region of the spectrum, as if it were emitted by an object that has a temperature of only a few degrees above absolute zero. In addition, because the material and energy in the early universe were uniform to a large degree, it is expected that the CMB will look the same in all directions.[9]

In 1964, motivated by his work on the CMB, Dicke convinced Princeton colleagues David Wilkinson and Peter Roll to construct a device capable of detecting the CMB experimentally. But, before the Princeton experiment was ready to go, Dicke got a call from an astronomer from Bell Labs named Arno Penzias inquiring about Dicke's work on the CMB. Penzias, it turned out, had a nagging problem with an antenna that he and a colleague, Robert Wilson, planned to use for radio astronomy. The antenna, a 6-meter structure in Holmdel, New Jersey, was originally built to detect faint radio waves bounced off satellites. Penzias and Wilson were refurbishing the antenna and studying it in order to make sure they could achieve the sensitivity needed for the astronomical observations they hoped to make. In the process, they discovered a constant tiny signal—what a physicist would call "noise"—that was not supposed to be present in the device. They struggled to determine the source of the noise, thinking it might be from a manmade source in nearby New York City or from a natural astronomical source in the galaxy. After ruling out these sources, Penzias and Wilson looked closer at the antenna itself, even going so far as to evict a pair of pigeons and clean up the "white material familiar to all city dwellers"[10] left behind by the birds. Nothing made the bothersome noise signal disappear. Upon complaining to a friend at MIT about the problem, Penzias was told about Dicke's work, which led Penzias to call Dicke and ask for the details. Shortly thereafter, Dicke and his coworkers traveled to Holmdel and, together, the Princeton and Bell Labs groups determined that Penzias and Wilson had unwittingly discovered the CMB.[11] Thirteen years later, Penzias and Wilson were awarded the 1978 Nobel Prize for Physics for the accidental discovery.

Many experiments made measurements of the CMB following its discovery and confirmed that it comes to us from all directions, and has the frequency distribution of a blackbody with a temperature of a few degrees above absolute zero—characteristics in full agreement with the hot big bang model. The CMB was also measured to be completely smooth within the sensitivity of the instruments used to make the observations. This posed a problem for the big bang proponents. Clearly we do not live in a smooth and homogeneous universe. There are galactic clusters and galaxies and stars and planets and hamsters. Any decent cosmological model must provide a way for this structure to originate. If the expanding universe was completely homogeneous and smooth at the time of recombination when the light of the CMB was emitted, how could the structure originate? If all the mass were smoothly distributed, where and why would gravity begin collapsing matter into the structures we see?

The missing piece fell in place in 1992 when a team analyzing data from a NASA satellite named the Cosmic Background Explorer, or COBE, announced the discovery of slight spatial variations in the frequency of the CMB across the sky. The variation is generally described as fluctuations in the temperature of the CMB, meaning it is a change in the frequency distribution corresponding to light coming from regions with very slightly different temperatures. The light emitted from slightly hotter, more dense regions is more energetic or bluer than the light emitted from the rarified regions.

The discovered structure was not very pronounced; the variations are only one part in 10^5. Imagine how unimpressed you would be if the weatherman said it was 70.0000 degrees in Albany and 70.0001 degrees in New York City. The difference doesn't seem like much. Still, these variations in the temperature of the early universe are enough to lead to the structure we see in our universe, and the pattern of the fluctuations is the key that makes the CMB a cosmological Rosetta Stone. The idea is that the non-uniformities in the big bang, as reflected in the CMB, led to regions with differing matter densities. Those density differences were amplified over time as the matter in the regions with high density collapsed due to the gravitational attraction.

It was a very exciting time when COBE announced the discovery of fluctuations in the CMB. Without some sort of structure in the CMB, cosmologists would be facing a major puzzle, and the big bang model would have a serious flaw. With the discovery, however, the structure became one the major pillars of evidence supporting the big bang theory. In addition, the quality of the data was very impressive. Before COBE, scientists studying cosmology weren't used to confronting detailed, unambiguous data. COBE changed that. In fact, many physicists think the COBE data marked the beginning of cosmology as a precise science.

The man who led off the presentation of the COBE results in 1992, George Smoot of UC Berkeley, put a voice to the enthusiasm and amazement many scientists felt at seeing the COBE results. Referring to the COBE photograph of the structure in the CMB, Smoot said, "If you're religious, it's like seeing God." Later he clarified what he meant in an interview with the *New York Times*, saying, "It really is like finding the driving mechanism for the universe, and isn't that what God is?... What matters is the science. I want to leave the religious implications to theologians and to each person, and let them see how the findings fit into their idea of the universe."[12]

Religious interpretation aside, the scientific community agreed heartily that the work was extremely significant. George Smoot and one of his COBE colleagues, John Mather at NASA's Goddard Space Flight Center, shared the 2006 Nobel Prize for Physics "for their discovery of the blackbody form and anisotropy [nonuniformity] of the cosmic microwave background radiation."[13]

COBE was followed up by another NASA satellite in 2001 called the Wilkinson Microwave Anisotropy Probe (WMAP) that has better sensitivity than COBE.[14] This increase in sensitivity has led to a more detailed map of the fluctuations in the CMB. WMAP measurements of the degree and size of the CMB fluctuations have been used to probe the evolution and makeup of our universe in ways that could not have been imagined a quarter century ago. The WMAP data have led to several very exciting results and insights that play a critical role in the ideas that are behind most of the cosmological multiverse concepts, as we will see in the rest of this chapter and the next.

One of the significant early results of WMAP was the determination of the geometrical curvature of the space-time of our universe. Recall from Chapter 2 that the equations of general relativity allow for space-time to have different curvatures and that the attractive nature of the gravitational force is seen as equivalent to the curvature of space-time. There is a continuum of possibilities for the way in which space-time is curved at a given point in space. One possibility is that the space has no curvature, which is called "flat space." Flat space corresponds to the Euclidean geometry we all know and love—or hate—from grade school days. In such space, planes look flat, like the floor of a basketball court, and two parallel lines drawn on a plane could be extended forever without crossing and the distance between them would be unchanging. Another possibility corresponds to what is called "positive curvature." According to Einstein, this is the type of curvature caused by mass (energy) that gives rise to the attractive nature of gravitation. In this type of space-time, planes curve back in on themselves and can be visualized as the surface of a sphere. In this geometry, parallel beams of light emitted in a plane will eventually cross. To see this, imagine drawing two lines on the surface of a globe parallel to the north-south direction. If the two lines are extended as far as possible, they will cross at the poles and each line will come back and join itself so that two circles are formed. In the case of negative curvature, a plane can be visualized as something like the surface of a saddle. Two parallel lines drawn on such a surface will diverge when extended. They will never cross, and the distance between them as measured from outside the plane will become larger and larger.

The space-time curvature at any particular point depends on the underlying geometry of the universe as well as the local environment. We know the universe must have dramatic variations in the curvature of the space-time within it as a reflection of the non-uniform distribution of matter. Galaxies and stars and planets and dogs all possess mass and must bend the space around them according to relativity. Objects with intense gravitational fields like stars and black holes create a much larger positive curvature in the adjoining space-time than does a dog. Still, these objects create only small, local perturbations in the large-scale curvature of the space-time of our universe. If you think of all these little, local distortions as being roughly evenly distributed and important only in their total contribution to the mass of the universe, you can imagine that the universe itself must have an overall, underlying curvature. The shape of the space in the universe depends on the total energy/mass density in the universe. Interestingly, there is a continuum of negative or positively curved possibilities, whereas flat space would be the exceptional case that would require the energy/mass density of the universe to have a very particular value, known as the "critical density."

To see how WMAP scientists managed to make a precise measurement of the geometry of the universe, we need to consider the nature and appearance of the structure in the CMB. If your eyes could see the CMB radiation and had the sensitivity of WMAP to the small color/temperature differences, the night sky would glow brightly, but it would appear to have a cosmic case of acne! The sky would seem splotchy with slightly different-colored regions of various sizes. The pattern of the different colored regions would appear quite random. But the pattern is not completely random, and it holds within it the answers to a number of very interesting questions about our universe.

In the very early stages of the big bang—say in the first trillionth of a second—there were non-uniformities in the energy density of the primordial soup of elementary particles that made up the universe at that time. The origin of those non-uniformities is not specified in the big bang model per se and will be a focus of the next chapter; for now, let's just assume the fluctuations in the energy density were present and those non-uniformities led to the slight variations in the density and temperature of the matter present in the universe at that time.

Although we may be speaking of something that happened long ago and occurred at energy densities not seen at a large scale since that time, a gas with variations in density and pressure is hardly something exotic. In everyday life we call such density variations sound waves! For example, a guitar string vibrates and slightly compresses, and rarifies the air molecules through its motion. That pattern of density changes propagates outward from the string at the speed of sound, and it is those density changes hitting our ear that we hear as sound. A loose, heavy string vibrates slowly. Because that slow vibration causes a low frequency of pressure changes in the air surrounding it, the frequency of the sound wave is low and it has a long wavelength. We hear the sound emanated from such a string as being a low pitch. Tight, fine strings, on the other hand, vibrate very fast and create waves with short wavelengths and high frequencies that we perceive as high pitches.

The non-uniformities in the early universe were regions of alternately compressed and rarified matter—sound waves in essence. Cosmologists who talk about the acoustics of the early universe are really studying the pattern of energy density fluctuations. It's more than a simple analogy; the patterns of compressed and rarified matter in the early universe propagated like sound waves, and the speed of propagation depended on the nature of the matter in the primordial soup and the way in the particles in that soup interacted with each other.

Cosmologists analyze the pattern of early universe fluctuations by looking at the so-called power spectrum, which is the degree to which the temperature changes for the different-size spots, which roughly correspond to the different wavelengths of the sound/density waves in the early universe. It turns out that the largest spots we see in the CMB, corresponding to the longest wavelengths of the density waves in the early universe, also show the largest variation in temperature. By knowing the speed with which the waves propagated in the early universe and the age of the universe at recombination when the CMB light was emitted, cosmologists can calculate the size of the regions in the early universe that correspond to these largest spots we see in the CMB. Knowing the true size of these regions, scientists can then compare this to a measurement of the *apparent* size of these largest spots to determine the large-scale geometry of the space in the universe. This is illustrated in Figure 6.2, where the apparent, or perceived, angular size of a spot is shown to depend on the geometrical path taken by the light as it traverses the 13.3 billion light years from point it was emitted at the time of recombination to the place where it is detected by WMAP scientists.

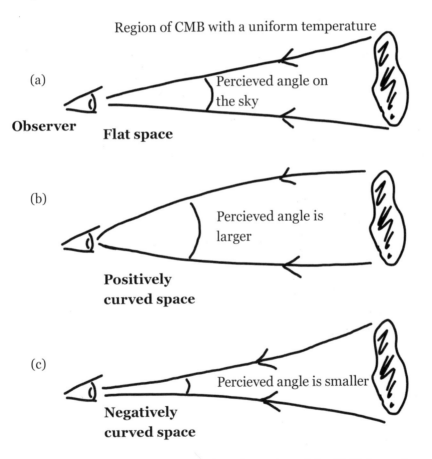

Figure 6.2: The apparent size of a region of the CMB depends on the curvature of the space in the universe.

Using the technique described previously, the WMAP team has measured the geometry of our universe to be flat, or at least very close to flat. This is an intriguing result, because flat space is a special case and science isn't built on the belief in coincidences. Physicists typically feel that special cases must happen for a reason. So, it's widely believed that it is likely that the flatness of the space we see in our universe is an important scientific clue. Because the big bang model doesn't have built within it a natural reason for the flatness of space, this clue indicates that big bang model is incomplete.

Actually, the fact that the space of our universe is flat leaves us with a much more intriguing problem than a scientifically implausible coincidence. In the equations of general relativity, the curvature of space depends on the energy/mass density of the universe. Recall that it requires a certain amount of energy/mass—a critical density—to make the space in the universe flat. Astronomers estimate that all of the normal matter and dark matter in the universe amounts to only 30 percent of the mass/energy required to make the space of the universe flat. That means that all the stuff we know about only gives us 30 percent of the critical density, yet the universe *is* flat!

Where is the rest of the energy? The answer is that we don't yet know. Because we know it must be there and we don't see it, physicists call it "dark energy." Whatever it is, dark energy constitutes 70 percent of the mass/energy of our universe.

The idea of dark energy isn't really new. In 1917, as he toyed with the equations of general relativity that describe the evolution of space in the universe, Einstein found that the universe in his model was prone to collapse due to the gravitational attraction of the parts within it. Because he thought that we live in a static universe, he added a term in his equations called the cosmological constant, which counteracted the force of gravity and stabilized the universe in his model. A few years later Alexander Friedmann discovered that general relativity is also happy with an expanding universe and no cosmological constant. Later, when the universe was found to be expanding, it seemed as if there might be no need for an ad hoc cosmological constant.

Needed or not, quantum field theory—unknown in 1917—provides a natural way for the cosmological constant to arise. Recall from Chapter 5 that the vacuum is "something rather than nothing" because of Heisenberg's uncertainty principle. With the amazing success of quantum field theory and the Standard Model, we know that virtual particle-antiparticle pairs are constantly popping in and out of the vacuum. This means that, at any given moment, the vacuum has a mass and an energy density. It happens that this energy density of the vacuum has exactly the correct form for creating a cosmological constant term in the equations of general relativity. It very well might be that dark energy—the missing 70 percent of the energy density needed to flatten the geometry of space—is the energy arising from quantum fluctuations in the vacuum. This makes for a tidy little picture that ties the large scale geometry of the universe to things happening at the quantum level. How cool is that?!

Unfortunately, there's a problem with the convenient, tidy little picture just painted. We know a great deal about the particles that exist in nature. After all, we have the Standard Model that works extraordinarily well in describing all sorts of strange and esoteric particle physics processes. When physicists use the machinery of quantum field theory, as well as what we know about the particles and forces of nature, to calculate the energy density of the vacuum, they come up with something that is about 10^{120} times larger than what we need to solve the dark energy problem. Yep, that's a 1 with 120 zeroes after it—an unimaginably large number. The cosmological constant derived from such a vacuum energy density is so large that the expansion of the universe would be vastly larger than what we see and the universe as we know it would not exist.

So, we have a big puzzle. The flatness of the universe tells us that we don't yet understand a large component of the universe's mass/energy budget. In principle, that missing 70 percent of the mass/energy of the universe might come from the energy in quantum fluctuations in the vacuum itself, except when that is calculated it is too big by 120 orders of magnitude! So, it seems that there is more to the quantum structure of the vacuum than we know about, and this may or may not be related intimately to the missing dark energy.

Is the vacuum energy problem a small issue or a big deal? Leonard Susskind, one of the founders of string theory, states in his wonderful, recent book entitled *The Cosmic Landscape*, that the vacuum energy problem "truly is the mother of all physics problems." I think that qualifies as a big deal.

The flatness of space and motherly problems aside, the big bang model is well supported by scientific evidence. The universe is seen to be expanding as mapped out by a universe of receding galaxies. The CMB has all the characteristics expected for the radiation released in the early universe when neutral atoms were formed. It has the correct spectral form and enough non-uniformity to account for the large-scale structure we see in the universe. The abundance of light elements agrees very well with expectations from big bang nucleosynthesis, where the bulk of the light elements in the universe were formed in the few minutes of time after protons and neutrons were formed and before the average energy dropped too low for nuclear fusion to take place. There are, of course, many questions that remain unanswered in the hot big bang theory, including the origin of the flat geometry. Much of this will be discussed at length in the next chapter. Still, with all this evidence supporting the big bang, it seems as though any cosmology that is to be taken serious scientifically must lead smoothly into a big bang, or at least create conditions that lead to the things that we perceive as having been caused by the big bang.

Many people have felt uncomfortable with the big bang concept through the years, partly because it implies a beginning to our universe—a beginning to time. A number of physicists, including Einstein, have speculated that what we witness as the big bang may be nothing more than one stage of an infinite series of "big bang–big crunch" cycles. The idea is that the expansion of the universe as we know it will slow down and eventually reverse due to the gravitational attraction of the matter/energy within it. In this scenario, the space-time of the universe will collapse someday to the point that some unspecified mechanism causes it to rebound into another expansion and contraction cycle. The collapse is often called the big crunch. The early proponents of this idea hoped that it would reinstate the eternal nature of our universe by replacing the single big bang with an endless series of big bangs. In this view you might say our universe is just one in an infinite series of temporally separated universes—a cyclic multiverse, if you will.

This big bang–big crunch cycle of universes is called the oscillating big bang multiverse in the populist taxonomy introduced in Chapter 1. This particular idea is no longer very popular among physicists, in part because there are technical difficulties in trying to construct a viable theory. The technical difficulties arise in two main areas. The first is that the magical moments between the big crunch of one cycle and the big bang of the next cycle necessarily involves conditions where the energy density reaches values where our current physical theories are inadequate. It is thought, for example, that quantum effects on space-time would be important at such an energy density, and, unfortunately, we don't yet have a quantum theory of gravity. If we assume this problem is going to be solved someday by some as-yet-undiscovered genius, there remains an issue with entropy and the second law of thermodynamics. According to the second law of thermodynamics, the entropy or disorder of an isolated system, such as a universe, will increase with time. That continues to be the case regardless of whether the universe is expanding or contracting. A physical chemist turned cosmologist at Caltech named Richard Tolman showed in 1934 that this entropy growth would causes the length of each cycle in a big bang-big crunch multiverse to be longer than the one before it.[15] Extrapolating back in time, the cycles become shorter and shorter until, eventually, there is a beginning. Without solving the entropy problem, the cyclic big bang scheme fails to create the endless universe that was the goal of the model in the first place.

Even if the technical problems were overcome, there is another important reason why the cyclic big bang multiverse advocated by Einstein and others in the 1930s is out of favor among physicists: It seems that the expansion of the universe is not slowing down, as you would expect for a big bang-big crunch scenario, but rather *the rate of expansion of the universe is growing*!

The stunning result that our universe is growing at an ever-increasing rate was announced in 1998 and 1999 by two teams of scientists, the High-z Supernova Search Team led by Brian Schmidt at the Australian National University's Mount Stromlo Observatory and Nicholas Suntzeff of Texas A&M University, and the Supernova Cosmology Project led by Saul Perlmutter at Lawrence Berkeley National Laboratory. These two teams basically played the same game that Hubble and Humason did in 1929, only they did it with more galaxies and with galaxies that are vastly more distant. Recall that Hubble and Humason compared the recession velocity of nearby galaxies as measured by redshifted spectra to the distance determined by looking at the brightness of a certain class of stars they could identify in the target galaxies. The astronomers who discovered the accelerating expansion in the late 1990s were able to look at much more distant galaxies by using the brightness of a certain class of supernova that is much, much brighter than the stars used by Hubble. So, they were able to look at much more distant galaxies and, consequently, measure the expansion rate of the universe far back in time and observe the change in the expansion rate as a function of time. They found, much to everyone's surprise, that the expansion rate has changed over time and that it is *growing*.

This newly discovered accelerating expansion of the universe adds to the puzzle surrounding the cosmological constant and the vacuum energy of our universe. The repulsive force that comes from dark energy has changed during the history of the universe. Perhaps some residual vacuum energy—once we understand why the vacuum energy is so small in the first place—will prove to provide the driving force increasing the expansion rate, though the fact that the vacuum energy is expected to be constant with time presents a serious problem for that solution. Some physicists have suggested that a new substance dubbed "quintessence," which permeates the universe and can evolve in time, could provide the needed repulsive force.[16] Experiments have been proposed to look at even more distant supernova—still further back in time—in order to determine the true nature of dark energy, as it is one of the most compelling physics problems of our time.

An accelerating expansion of the universe kills the oscillating big bang multiverse outright. It seems as if there will never be a big crunch as conceived in the 1930s. Still, what kills one idea enables another. For example, Lauris Baum and Paul Frampton at the University of North Carolina, Chapel Hill, suggested in 2006 that the existence of dark energy might provide a way around the entropy issue that has plagued cyclic cosmologies.[17] The idea is that dark energy causes the universe to expand at an ever-increasing rate. In fact, the expansion rate can grow so much that the edges of the universe recede faster than the speed of light. This would mean that the observable universe of the future encompasses less and less of the energy/mass currently in our observable universe. As time passes, all the matter in the universe gets spread out so that each little bit is far out of causal contact with all the other bits. Within each of these causal patches of the hugely expanded reality, the entropy would be effectively zero. Now, within the model something happens (the details of which I'll happily ignore here) that causes the expansion to stop and the space in each individual causally disconnected region begins to collapse independently. Each of these patches independently does a big crunch and bounces back into something like the big bang—complete with dark energy–laden space expanding—and the cycle repeats indefinitely into the past and into the future. In the big bang for each individual patch,[18] the entropy is reset but it gets zeroed out again in the succeeding expansion. So, the cycle can repeat indefinitely. In one cycle, a single "universe" fragments into a huge number of independent "universes" forming what I call the "cyclic patch multiverse."

There is nothing fundamental in the big bang model that limits the size of the greater reality relative to what we can see. In fact, current cosmological/astronomical evidence implies that the greater reality of space is vastly larger than our observable universe and even much larger than our cosmological event horizon.[19] Recall that the observable universe is that portion of the universe where light has had time to reach us.[20] That would be a sphere with a radius of about 13.7 billion light years centered about us. The cosmological event horizon represents the point beyond which we could never see *even if we wait forever*. This may seem confusing at first because if we wait forever you might expect the light from any point in the great reality, no matter how distant, to have time to travel to us. But remember that the space in the universe—and the greater reality—is expanding and that points that are farther away are moving away from us faster. So, moving directly away from us out into the greater reality, there is a point where the space is expanding away from us faster than light can travel. In other words the expansion carries the space away from us faster than the edge of the observable universe is growing. In the words of Arizona State University Cosmologist Lawrence Krauss, "Nothing can move through space faster than the speed of light, but space can do whatever the hell it wants as far as we know."[21] In fact, because the expansion rate of the universe is growing, this effect will become more and more important causing the wonderful sky full of galaxies we see today to move outside of our observable universe one day.[22]

Why is it that scientists feel there is more reality out there than we'll ever be able to see? Well, space is flat, or very close to flat, and the expansion of the universe is accelerating. Flat space means the space in the universe does not fold back in on itself, as you would see in a universe with positively curved space. If we head away from Earth in a straight line and travel forever, we won't find ourselves coming back to Earth from the opposite direction. In spite of the flat space, you can imagine a topologically bounded space. The universe could operate like a room where you walk through the door on one side of the room only to find yourself walking in the same room from the other side. People have searched for this by looking at the repetition of patterns in the CMB and have found no evidence for this so far to my knowledge. In addition to the flat, expanding, seemingly not bounded space, the universe looks the same on large scales in all directions—what a physicist would call "isotropic"—and it is thought to look the same on large scales no matter where one stands in the universe. The implication of this is that what we see continues well beyond the region in which we could ever possibly see it, as limited by the finite speed of light.

If the greater reality extends far beyond what we can see and beyond what is causally connected to us, we can think of that great reality as consisting of different causally disconnected regions, each of which would seem much like our universe[23] in terms of overall structure and physical laws. Only the initial conditions of each of the regions—the structure—would differ, because the pattern of the fluctuations in the greater reality would vary from place to place. Things would look much the same, but the details would differ. This concept is what I have dubbed the "beyond-the-horizon multiverse." It corresponds to Max Tegmark's Level I multiverse.[24]

There is a little twist to add to the concept of the beyond-the-horizon multiverse. There are compelling scientific reasons to believe that a process very early in the big bang took something very small and blew it up almost instantaneously into something very, very big—vastly larger than the observable universe. Physicists call this phenomenon "inflation," perhaps one of the most extreme cases of scientific understatement in a name ever to be put forth. Inflation is the main focus of the next chapter. Once we have that under control, we'll return to the beyond-the-horizon multiverse.

7
The Ultimate Free Lunch

The big bang theory enjoys widespread acceptance. Scientists no longer argue about whether or not the universe is expanding. Instead, the issues center around the evolution of the rate of expansion. Are we really expanding faster and faster? What was the expansion rate like when the universe was younger? The CMB radiation is acknowledged to have the exact form expected to come from the big bang. The discussion in scientific circles concerns what we can learn from the patterns of the tiny fluctuations in the CMB and what those fluctuations can tell us about the first moments of the big bang. There is general acceptance that the model of light element synthesis in the big bang—big bang nucleosynthesis—leads to the correct abundances of light elements in intergalactic space. Incredibly, the big bang model successfully accounts for *all* these complex phenomena in detail, and that makes for a very compelling case scientifically.

Still, in spite of all its amazing success, the big bang model cannot stand alone. Taken by itself, the theory has some serious problems. One problem with the big bang is that, extrapolating some 13.7 billion years back in time, the universe collapses to a point and the energy density of the universe becomes infinite. Physical principles as we know them are applicable back at the tiniest fraction of a second after the start of the big bang, but prior to that the details of the big bang are sketchy. In the words of Alan Guth, an MIT physicist who will figure prominently in this chapter, the theory "makes no attempt to describe what 'banged,' how it 'banged,' or what caused it to 'bang.'"[1]

The energy of particle collisions taking place at the Large Hadron Collider corresponds to the typical collision energy of particles in the early, hot universe at approximately 10^{-15} seconds after the start of the big bang. Extrapolating back before 10^{-15} seconds we lose the direct knowledge gained from controlled laboratory experiments. That's not to say we can't infer things from what we can see at lower energy or in astrophysical observations; it's just that the ice under us becomes thinner and more prone to give us a surprise as we consider events further back in time.

Even without a big surprise, as we go back in time, the energy of particles becomes very large, and that means their gravitational effects become large. Eventually we hit the point, called the "Planck scale" by physicists, where the gravitational interaction between particles rivals that of the other forces. At the Planck scale of energy, the quantum wavelength of a particle is about the same as the size of the distortions of the space about the particle due to gravity. To understand processes at the Planck scale, we need a theory of quantum gravity, which is not something we have in our theoretical toolbox yet.

The flatness of the space in our universe also poses a problem for the big bang picture. There's nothing in the theory that *prohibits* the space from being flat. Rather, the problem is that flat space is a very special case, and there's nothing in the theory to cause it to arise naturally. Physicists, and scientists in general, abhor coincidences. Surely any complete cosmology will have a compelling reason for the space to be flat.

The big bang theory also encounters difficulties arising from the large-scale structure of the universe. This may sound odd, because the tiny fluctuations in the CMB radiation lead to the large-scale structure and the CMB is a core prediction of the big bang. The problem is that the big bang model, on its own, does not provide a good reason for those fluctuations in the CMB to be present. What is it that causes the structure in the first place?

Ironically, it's the *lack* of structure—the incredible uniformity of our universe—that presents one of the most serious problems with the big bang theory. After all this discussion about the fluctuations in the CMB, it's important to recall that the variation is only one part in 100,000. For this degree of uniformity to be present, the early universe must have been all at the same temperature to one part in 100,000. In scientist lingo, the early universe was in thermal equilibrium. Let's think this through. We see the photons that make up the CMB coming to us from all directions. These photons were emitted around 380,000 years after the start of the big bang. This means that any bit of space in the universe at that time was causally connected to a region of space surrounding it with a radius of 380,000 light years. Since that time, the universe has expanded and the observable universe has increased in size. Because the speed of light is greater than the rate of expansion of space during this time, the observable universe today encompasses regions of the universe that had not been in causal contact at the time the CMB photons began their journey to us. If this is true, how could the CMB seen today on one side of the sky be at the same temperature as the CMB on the other side of the sky? In the big bang model, the CMB photons from these two widely separated regions are emitted from places that were so far apart at 380,000 years after the big bang that the exchange of information or heat energy could not have happened. Hot and cold spots could not have been smoothed out over such a scale. This is known among cosmologists as the "horizon problem," and it is illustrated in Figure 7.1 on the following page.

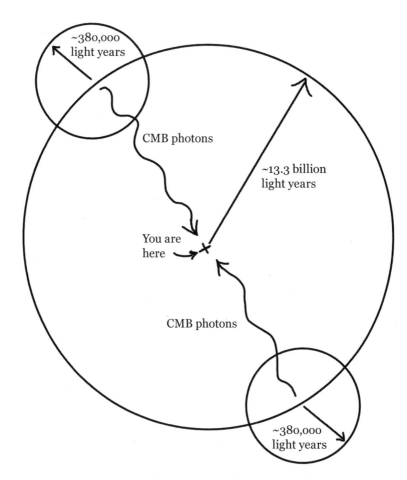

Figure 7.1: Cosmic microwave background light reaches us from regions of the universe that had never been in causal contact at the time the photons began their journey, yet the data indicate that these photons were emitted from regions with the same temperature to one part in 100,000.

Let's consider a silly example as an illustration of the horizon problem. Imagine that the outside of a tall apartment building with 1,000 apartments is to be painted. The painting will be done by 1,000 painters. Each painter is assigned the responsibility of painting the balcony of one of the 1,000 apartments. The painters are each provided with cans filled with paints of primary colors and a bucket in which they can mix paint. With no communication before or during the process, at an agreed-upon time each painter mixes paint and paints his assigned balcony. Afterward a careful inspection of the building shows that the colors of all the balconies agree to within one part in 100,000. *That's* the horizon problem.

Of course, such a thing with the balconies and the paint can't *possibly* happen without some sort of advance planning. That's the catch with the horizon problem in the big bang: There is no mechanism for advance planning in the big bang. The CMB photons we see provide a record of the temperature in parts of the universe so far flung that at the time the CMB was formed there had not yet been time for thermal equilibrium to have been established—or even for light to travel—between all the points we see today. Yet the temperatures of these far-flung regions are the same.

Whether it makes sense naively or not, the fact is that the CMB is uniform to a part in 100,000. So, the horizon problem means either we throw out causality and the limiting speed of light, or we admit that the big bang model needs help. Combining the horizon problem with the other issues—problems with infinite energy density as we approach t=0, lack of a natural reason for flat space, and the lack of a way for the fluctuations in the CMB to arise (plus at least one more good reason[2] I'm omitting here to avoid the technical discussion)—it seems as if the big bang theory is lacking some critically important component.

What a dilemma! The data tell us the big bang had to happen, and yet there are serious problems with the big bang concept that must be addressed before the model can be taken too seriously. As it happens, physicists have thought of a way out of this quandary, but the proposed solution to this dilemma is very strange.

Remember the old adage that you should be careful what you wish for? It brings to mind a joke I heard once. Two men died and went to Heaven. Upon greeting them, St. Peter said, "I'm sorry, gentlemen. It seems we've had some construction issues with your heavenly housing that will delay your entry here. Until we have a place for you, I'll send you back to Earth in whatever form you wish."

Upon hearing this, the two men were quite excited. "Great!" said the first guy. "I want to be an eagle soaring above beautiful country!"

"No problem," replied St. Peter. POOF! The guy was gone. "And what do you want to be?" St. Peter asked the other guy.

"I'd like to be one cool stud!" was the reply.

"Fine," replied St. Peter, and the other guy disappeared.

A few weeks later, St. Peter learned the housing for the two men was ready, and he sent an angel to bring them back to Heaven. "You'll find them easily enough," he told the angel. "One of them is soaring above the Grand Canyon, and the other one is on a snow tire somewhere in Detroit!"

Okay, the solution to the big bang dilemma is probably not as bad as being a cool stud in Detroit. But it *is* a mind-bender. The fix for the problems with the big bang is called "inflation," and theories that incorporate this fix are called "inflationary big bang models."

The concept of cosmological inflation is rather simple, albeit a bit wacky if you've not heard it described before. The idea is that what became our universe started out very tiny, perhaps as a quantum fluctuation. In *what* it fluctuated, we'll ignore for now, but it could be that reality at that stage was vastly larger than the fluctuation that becomes our universe. At this time, the fluctuation that became our universe was all causally connected because it was so very tiny—perhaps only 10^{-50} meters in radius.[3] Then, after about 10^{-36} seconds, the universe underwent a period of inflation where it got *very* big *very* fast. In fact, the term *inflation* is a rather dramatic understatement for what happened in this picture. There was an unimaginable, exponential growth—by at least a factor of something like 10^{50}—in the time between 10^{-36} seconds and 10^{-35} seconds. In case the scientific notation sanitizes things a bit much, let's look at it without the powers of ten. The growth was by a factor of 100,000,000,000,000, 000,000,000,000,000,000,000,000,000,000,000,000 in 0.0000 0000000000000000000000000000001 second! Growth that is *far* too much to comprehend happened in a time that was so short as to be completely unimaginable. The expansion happened much faster than the speed of light. Relativity is okay with that because it is the space that expanded rather than a signal of some sort being passed around.

The numbers may vary a bit from theorist to theorist, but the basic idea is that what was to become our universe started out very tiny and, in the briefest of instants, became vastly larger than the observable universe at that time. After this moment of inflationary expansion, the energy driving the expansion was dumped into the radiation and subatomic soup of particles that were the constituents of the big bang at that time. From that stage forward the inflationary picture follows the big bang scenario.

The idea of inflation may seem sort of crazy, but it does a very nice job of solving the problems with the big bang theory. Consider the flatness problem, for example. Because flat space is a special case, let's suppose the universe begins very small with some arbi-trary non-flat curvature to the space—some sort of curvature that can't be considered anomalous or special in the eyes of a cosmolo-gist or general relativist. If this space is blown up to be sufficiently large by inflation, it will seem flat. To see this, imagine a baseball, with obvious curvature, blown up to be the size of the Earth, or even much larger. Such a huge baseball would seem quite flat locally. The more the inflation, the flatter the space. In fact, inflation drives any geometry toward flatness. The flatness of space is a natural outcome of inflation.

How about the horizon problem? Does inflation help with that? You bet it does. Prior to inflation, what became our universe was very small and causally connected. In this picture, before inflation, it was small enough to be in thermal equilibrium. Through infla-tion, regions of space that were once in causal contact and thermal equilibrium were separated by such distances that they now seem as if they must have always been outside of causal contact. Yet, because they really were at the same temperature prior to being separated by inflation, they were at the same temperature just after inflation, and we now perceive the CMB photons coming from these far-flung regions to have the same blackbody spectrum. Because everything in our observable universe began from the same tiny thing, the incredible degree of uniformity is to be expected.

Naively, it might seem that, because that which became our observable universe started out very small and in thermal equilibrium, things must be completely uniform. But that's not true. Quantum fluctuations come in different sizes and at all times. There are quantum fluctuations within quantum fluctuations. So, the idea in the inflationary picture is that in the tiny volume that inflated to become our universe there were small quantum fluctuations in energy that, when blown up to cosmic scales, were frozen in place because of how inflation spread things beyond causal connectedness. Fluctuations happened during the time of inflation as well, leading to small non-uniformities in energy scattered throughout the universe with a spectrum of spatial sizes. Within the inflationary big bang model, this is the origin of the structure in the CMB, which led eventually to the large-scale structure in our universe. String theorist and popular science writer Brian Greene summed it up beautifully with a statement that I always present to my classes with a grand, dramatic flair: "The more than 100 billion galaxies, sparkling throughout space like heavenly diamonds, are nothing but quantum mechanics writ large across the sky."[4]

The idea of inflation entered the mainstream physics community in 1980. A postdoc at the Stanford Linear Accelerator Center, Alan Guth, stumbled upon the idea while trying to solve the so-called "magnetic monopole" problem.[5] Magnetic monopoles are strange particles (I'll spare you the details) that we don't see in the universe today. Yet, the best estimates in 1980 led to the expectation that a great many of them should have been produced in the earliest stages of the big bang. Alan Guth realized that he could work around this problem if something inflated the space in the universe dramatically after the monopoles were formed reducing their number in the observable universe to a level that could not be measured. Then he realized inflation solved both the horizon and flatness problems in cosmology, as well as the magnetic monopole problem, and inflation was born.

As is often the case with big scientific breakthroughs, the fruit had ripened and others were close to picking it. In particular, a Russian named Alexei Starobinski, at the L.D. Landau Institute of Theoretical Physics in Moscow, had put forth a realistic model of inflation a year before Guth hoping to avoid the infinite density at a point—the so-called singularity—at the beginning of the big bang. However, his work didn't mention the horizon and flatness problems, and it did not get much attention.[6]

Guth's work, however, did get attention and by 1982 cosmologists around the world were thinking about the implications and details of inflation. For example, inflation, once begun, was difficult to stop in Guth's model. Important modifications to the theory of inflation were put forth by Andrei Linde and independently by Paul Steinhardt and Andreas Albrecht in 1982, which solved this *graceful exit* problem. Also in 1982, four separate groups attending a workshop at Cambridge University[7] worked furiously to calculate the energy density fluctuations in the inflationary picture. By the end of the workshop, these physicists, including Stephen Hawking, Alexei Starobinski, Alan Guth, So-Young Pi, James Bardeen, Michael Turner, and Paul Steinhardt, converged on the theoretical expectation for the inflationary density fluctuations.

The work at the Cambridge workshop was a critically important development because, according to the theory, these fluctuations are thought to have led to the large-scale structure in the universe. At the time, the nature of these fluctuations in the real universe had not been measured. Since then, of course, the spectrum of fluctuations in the CMB has been observed, and it agrees well with the expectation from inflation! (More on this later.)

Though inflation is a convenient trick that seems to solve many problems associated with the big bang, we are left with questions as to how and why the universe might undergo such an inflationary period. A number of ideas have been proposed. In fact, it could be that the mechanism that drove inflation—assuming it happened—is similar to dark energy, whatever that is. After all, in both cases, there is an expansion of the space. Though it may well be that the two phenomena have different causes. No one knows yet.

The theoretical discussion about the cause of inflation centers around a hypothetical "inflaton" field. Fields are nothing new in physics, of course. Earlier, we encountered electric and magnetic fields, which are central to our understanding of the electromagnetic force and light. Recall that a field is a mapping of some characteristic to all points in space. In the case of the electric field, for example, the field is a measure of the magnitude (strength) and direction of the electric force encountered by a positive charge were it to be placed at that point in space. Fields don't always need to have both strength and direction. Sometimes they only have strength. A field that is a mapping of the temperature of every point in space is an example of a field with no direction. Fields having only a magnitude are known as scalar fields. The inflaton field is thought to be a scalar field.

Particles are described as fields in quantum field theory. The idea that something highly localized like a particle can be described in terms of a quantity that is spread over all space may seem strange at first, but that's the nature of quantum mechanics where particles are thought of as waves. The fact that the field extends over all space does not mean that its value is nonzero everywhere. For example, the electric field of any particular electromagnetic wave is modeled as if it extends over all space, but the value of that field is effectively zero over the vast majority of the universe. It is only nonzero in the region where that wave is found. In a region of space where there is nothing— the vacuum—the energy of a field is at a minimum. A physicist would say the potential energy stored in the field, or just "the potential" of the field is at a minimum.

The concept of potential energy is an important one in both classical and quantum physics, and it turns out to be very important for understanding mechanisms of inflation. As silly as it sounds, a rolling marble is a great example for visualizing potential energy and how it relates to fields in particle physics and cosmology. Imagine a marble inside a curved bowl. As long as the marble is confined to the bowl, it has a minimum of energy when it is sitting motionless at the bottom of the bowl. If the marble is located at some point up one of the curved sides of the bowl, momentarily motionless, a physicist would say it has "potential energy," energy stored in it by virtue of its height. If released while up on that wall, the marble will begin to roll down toward the center of the bowl, converting the potential energy into the kinetic energy of motion. If the marble starts higher up on the wall of the bowl, it will be moving faster at the bottom. This means that the energy stored in the marble is larger if it is positioned higher on the wall of the bowl. The sketch in Figure 7.2 shows how the potential energy stored in the marble depends on where the marble is within the bowl.

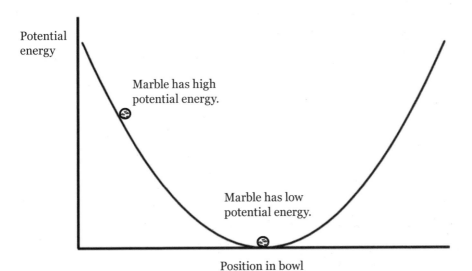

Figure 7.2: The gravitational potential energy stored in the marble is a function of its position in the bowl. The higher it is, the larger the potential energy.

None of this is really new. Most of us grow up realizing that, if given a choice in the matter, we'd rather have our toes under something dropped from a small height than the same something dropped from a large height. A physicist would say the object has more potential energy stored at the larger height. This potential energy is converted to kinetic energy as the object drops toward your toes. The higher the object when it begins to fall, the greater its kinetic energy when it smashes your toes.

In quantum field theory, a scalar field can take on different values, and the energy density of the field in the vacuum depends on the value of the field. We can visualize the energy density of the field in the vacuum and how it depends on the value of the field, much like we visualize the way the potential energy stored in the marble in the bowl depends on its position, which in turn determines its height. Typically the value of the field is at a minimum in much the same way the potential energy stored in the marble is at a minimum when it sits at the bottom of the bowl. Physicists call the value of the field where the field's energy density is at a minimum "the true vacuum value of the field."

Because they are quantum mechanical beasts, the value of scalar particle fields can fluctuate from point to point and over time. It is *quantum* field theory, after all. Typically the field fluctuates away from the true value of the vacuum by a little and moves back in much the same way the marble will roll back down to the bottom of the bowl if it is displaced a bit and released. Exactly how the energy density of the field depends on the magnitude of the field is a defining characteristic of the field. Again, this is not so different for the marble. As illustrated in Figure 7.3, it will take much more effort to displace the marble by a certain amount from the center of the bowl for the potential curve A than it does for the potential curve B. The same kind of thing can happen for fields. The energy density of the field can vary as the magnitude of the field changes.

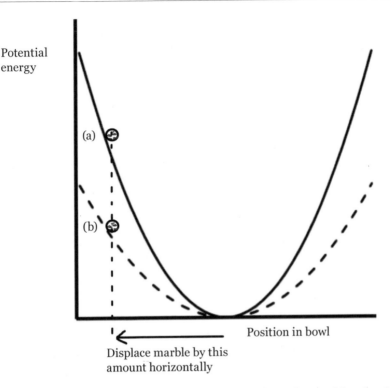

Figure 7.3: *It takes more energy to displace the marble a horizontal distance from the center of the bowl for a steep bowl (a) than for a shallow bowl (b).*

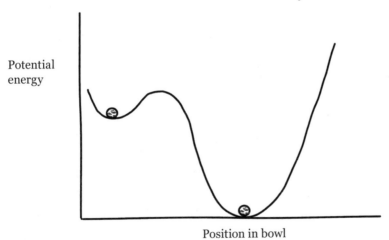

Figure 7.4: *The marble settles stably in both positions. However, it has a nonzero potential energy in the higher position.*

You could imagine creating a bowl with a dip in the lip of the bowl on one side. A cross-section of the bowl I have in mind is sketched in Figure 7.4. In this case, as in Figure 7.2, there is an absolute minimum potential energy position for the marble at the bottom of the bowl. There is also the dip up on the lip where the marble can rest where the potential energy is not at a minimum. It is possible to imagine a scalar field where the way in which the energy density in the vacuum varies with the field magnitude looks something like Figure 7.4. In such a case, if the scalar field settles in the higher dip, it is stable. At least, it is stable until a quantum fluctuation in its value allows it to tunnel out of the dip into the main part of the bowl, and the probability of that happening depends on the details of the height and width of the hill between the dip. When the field settles in the higher dip and has a nonzero energy density in the vacuum, the space is said to be in a "false vacuum."

For our purposes, the important thing about a scalar field in a false vacuum is that it contributes to Einstein's cosmological equations in the same way as does the cosmological constant, which is to say, it creates a term that provides a repulsive force seeking to expand the space. This means that a scalar field in a false vacuum state—the hypothetical inflaton field—can provide the mechanism to drive inflation.

If the inflaton field started up in the dip of a potential like that shown in Figure 7.4, the energy density in the vacuum would be high and the cosmological constant would be high and inflation would go on and on. One way to have inflation last a very short time before it ceases is to have the inflaton potential look something like Figure 7.5. In this case the inflaton field starts out with a high energy density (up on the plateau) and stays there for a time before it rolls off the edge and rapidly descends to the low energy density of the true vacuum. If the inflaton field was formed up on the high plateau in Figure 7.5, it would drive inflation while slowly moving down the roughly flat part of the potential. According to this picture, when the field reached the sharp drop-off, it decayed to the true vacuum very quickly and the big bang was born.

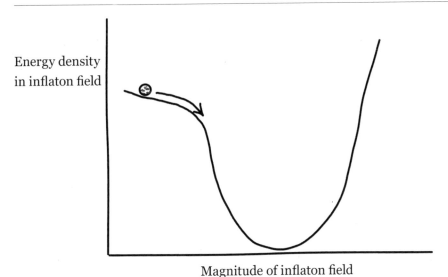

Energy density in inflaton field

Magnitude of inflaton field

Figure 7.5: This a hypothetical curve showing the energy in the inflaton field as a function of the value of the field. In this case, the field starts with a high energy density and drops slowly, causing inflation to happen in the universe during that time, until it reaches the steep slope and drops quickly to the minimum energy density with no inflation.

Naively, it seems that inflation *must* violate energy conservation. From something very tiny springs a reality vastly larger than the observable universe in less than a heartbeat. Surely that must take an unimaginably large amount of energy. It's hard to imagine the quantum trickery of Heisenberg's uncertainty principle working for this case.

The way around energy conservation in the inflationary picture lies in the balance between the energy stored in the inflaton field and the gravitational energy in the universe.[8] As the space expands during inflation, there is more and more energy stored in the inflaton field. This happens because the inflaton field has a positive energy density and the volume of space in which that field is found is increasing. If you have a larger volume of something with the same energy density, there is more energy overall. Recall, this is the energy that is dumped into mass and radiation when the inflaton field decays from the false vacuum to the true vacuum at the end of inflation. So, it's fair to say the energy stored in the inflaton field is what became the matter and energy in our universe as we perceive it. As the space—and the additional volume of the inflaton field—is created during inflation, the gravitational field throughout that volume is also created. At the end of inflation, for example, the universe is filled with matter and radiation and each bit of that stuff attracts all the other stuff. The way around the energy conservation problem is that the energy stored in that gravitational field is negative.[9] This is a technical detail that is as true in high school Newtonian physics as it is in advanced cosmology. During inflation the positive energy of the expanding space and inflaton field is balanced by the increasing amount of negative gravitational energy. According to Alan Guth:

> The total energy—matter plus gravitational—remains constant and very small, and could even be exactly zero. Conservation of energy places no limit on how much the universe can inflate, as there is no limit to the amount of negative energy that can be stored in the gravitational field.... If inflation is right, everything can be created from nothing, or at least from very little. If inflation is right, the universe can properly be called the ultimate free lunch.[10]

Through inflation, the universe expanded to a size that was vastly greater than the observable universe. Inflationary cosmology says there are regions of the greater reality that were causally connected initially and thermalized along with the region we inhabit that became causally disconnected through inflation. This process can create a huge beyond-the-horizon multiverse (Tegmark's Level I multiverse).

If you are ready for a little weirdness, let's consider the size of the beyond-the-horizon multiverse. According to the cosmologists Alexander Vilenkin of Tufts University and Jaume Garriga of Tufts and the Universitat Autonoma de Barcelona, it is a "generic prediction of inflation" that the space that was thermalized initially with our region of the unverse is spatially infinite.[11] If we consider the size of our observable universe as sort of volume unit of measure within the beyond-the-horizon multiverse, there must be an extremely large number of regions similar to our observable universe that are causally disconnected from us and from each other.

Admittedly, this universe counting may seem a bit confusing. The boundaries between the regions are not crisp the way it would be if they were like marbles filling a bucket. Instead it is more like marble-sized regions of sand in the bucket. It isn't so clear where one marble-sized sand region begins and another ends. Still, it does make sense to ask if two points in the bucket are more than a marble's width from each other and it makes sense to ask how many marble-sized regions of sand exist in the bucket. If inflation is true, our greater reality bucket contains an extremely large number of observable universe-sized regions.

Within all regions of the greater reality that, at the beginning of inflation, were causally connected and thermalized with the space that became our observable universe, we expect the basic characteristics of the vacuum to be similar. This means that all regions of the beyond-the-horizon multiverse have similar quantum fields and particles and forces—and the same basic physical laws and physical constants. However, the random quantum processes that led to the density fluctuations in each part of the beyond-the-horizon multiverse were different, meaning that the various regions had differing initial conditions. Recall that it's these fluctuations that determine the structure in our observable universe, such as galaxies and planets and dogs. Similarly, the differing fluctuations in the other regions of the beyond-the-horizon multiverse led to the galaxies and planets and whatever else may be out *there*.

Garriga and Vilenkin have determined that, if inflation happened, there are an infinite number of regions within the beyond-the-horizon multiverse that are the size of our observable universe and that all the possibilities that could happen due to differing initial conditions is finite.[12] In other words, there are only so many things that *could* happen and an infinite number of places for them to happen. Max Tegmark says that inflation generates all possible initial conditions with a nonzero probability. He adds that "... pretty much all imaginable matter configurations occur in some [region] far away, and also that we should expect our own [region] to be a fairly typical one."[13]

The mind-boggling implications of all this are rather well disguised by the scientific lingo. According to Tegmark, "...it means that everything that could in principle have happened here did in fact happen somewhere else."[14] There is a universe within the beyond-the-horizon multiverse where what *could* happen, *did* happen—including one where there is an identical you reading an identical book. Tegmark estimates the closest identical you is about 10 raised to the power of 10^{29} meters away—an unimaginably huge distance. He suggests that there is a region of space that is identical to ours within 10 raised to the power of 10^{115} meters of us. These are not precisely known numbers, it seems, but they are based on an accounting of the quantum possibilities and the size of regions over which things must vary in this picture. They are inferences based on scientific reasoning using known principles of physics.

The idea that anything that *could* happen, *does* happen somewhere leads to some very strange things. There are an infinite number of regions that share the same history but do not necessarily share the same future. Garriga and Vilenkin tell us, "Some readers will be pleased to know there are infinitely many [regions of the beyond-the-horizon multiverse] where Al Gore is President and—yes—Elvis is still alive."[15]

Okay. Take a deep breath. Yes, it is true that I just said a scientifically supported concept of the multiverse contains universes where Elvis lives. "Scientifically supported" here does not mean we have direct evidence of Elvis being all shook up in some other universe. The flatness of space and the uniformity of our observable universe make it reasonable to infer that the great reality of space is much larger than our observable universe. The concept of inflation, which is a core part of the scientific cosmological paradigm today (known as the "concordance model"), fixes the problems with the big bang and makes testable predictions. Inflation, if it happened, would also produce the huge beyond-the-horizon multiverse that I've been describing. The existence of the beyond-the-horizon multiverse and its characteristics are inferred to exist from things that are based on solid science, which is why I categorize it as a scientific multiverse concept.

Now, as if the beyond-the-horizon multiverse isn't strange enough, shortly after the 1982 workshop at Cambridge where cosmologists converged on the expectation for inflation-driven density fluctuations, cosmologists Andrei Linde and Alex Vilenkin and others realized that once started, inflation is eternal.[16] Figure 7.6 on the following page illustrates a couple of different possibilities for how this might work. Imagine the inflaton energy density in the vacuum varies in the way shown in Figure 7.6 (a). In this case, if the inflaton field settles at point "i", there is a false vacuum—the energy density of the inflaton field in the vacuum is high—and inflation happens. During inflation there is tremendous expansion of the space except where, here and there, a fluctuation in the inflaton field causes it to be in a region on the hill such as point "ii", where the field rapidly relaxes to the true vacuum. In these little regions where inflation stops, the energy stored in the inflaton field is dumped into radiation and particles and a big bang begins. In the meantime, inflation continues elsewhere. In this picture, the greater reality consists of causally disconnected bubble, or pocket, universes scattered randomly within an eternally inflating matrix. When the inflaton field is in a region where the energy density is falling rapidly, such as shown in Figure 7.6 (b) or at point "ii" in Figure 7.6 (a), quantum fluctuations can happen in either direction. A fluctuation taking the inflaton field up the hill causes inflation to happen faster in that spot while fluctuations taking the inflaton field down the hill slows things down. Again, the end result is a countless number of bubble universes in an eternally inflating matrix.

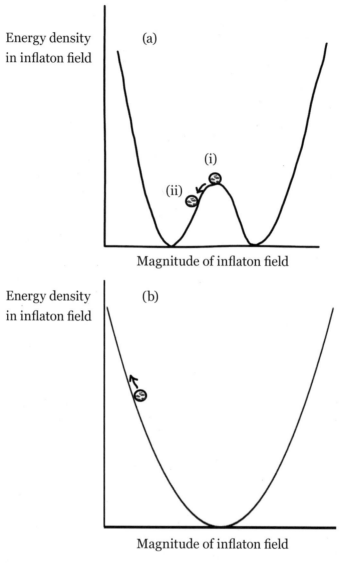

Energy density
in inflaton field

(a)

(i)

(ii)

Magnitude of inflaton field

Energy density
in inflaton field

(b)

Magnitude of inflaton field

*Figure 7.6: Inflaton potential curves leading to eternal inflation. In (a),
the inflaton field is in a stable postion at (i), but a quantum fluctuation in
the magnitude of the field will cause it to fall into one of the valleys of the potential (ii), causing inflation to stop at that point. In (b), the inflaton field is
in a region where it is falling to a lower energy density stopping inflation;
but, in some places the inflaton field will fluctuate upward in energy density
causing enhanced inflation. Either way, there are regions of space where
inflation stops and regions where it continues, making for eternal inflation.*

In Appendix A and Chapter 1, I categorized this reality of bubble universes embedded in a matrix of inflating space as a scientifically-supported, space-time separated multiverse. I call it the "bubble multiverse." Tegmark categorizes this as a level II multiverse.

Each bubble within the bubble multiverse springs from a single causally connected region of the inflating matrix where the inflaton field fluctuated and relaxed into the true vacuum. Though there must be some dependence on the details of how the inflaton field shifts to the true vacuum, each bubble universe within the bubble multiverse could contain within it a beyond-the-horizon multiverse. In Tegmark-speak, the level II multiverse contains within it a countless number of level I multiverses. This is illustrated schematically in Figure 7.7.

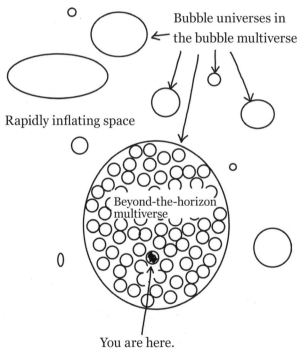

Figure 7.7: The bubble multiverse consists of regions where inflation has stopped in a matrix of eternally inflating space. Bubble regions where inflation has stopped might be much larger than the size of causally connected regions within the bubble, leading to a beyond-the-horizon multiverse. The bubble multiverse consists of multiverses within a multiverse.

Within the bubble multiverse, each bubble comes from a different place in the inflating matrix where the inflaton field fluctuated and relaxed to the true vacuum. The other physical constants of nature were also fixed at this time. Differences in how these constants settled down at the fluctuating point in space—or, if you are a believer in string theory, differences in the way the hidden dimensions are organized and shaped (more on this in Chapter 8)—lead to universes with different physical laws and different particles and particle masses.[17] Physicists estimate there might be as many as 10^{500} different unique possibilities for the set of physical parameters, leading to a countless number of different types of universes. Though we might expect our universe to be typical in some sense, many of these universes can be very different from ours. It doesn't take much of a change in the physical parameters describing the forces and particles in a universe to create something very different from what we see in our universe. For example, in the first few minutes after the big bang, if the weak nuclear force was slightly stronger, neutrons would have decayed very quickly in the first minute after they were formed and the universe would have been made entirely of hydrogen. If the weak nuclear force was slightly weaker, few neutrons would have decayed in the early universe and all the protons and neutrons would have been bound together into deuterium. In either case, the stellar synthesis of elements essential to life as we know it would not have happened.[18]

Quantum fluctuations in the inflating space-time matrix lead to what is practically an infinite number of different types of bubble universes. Quantum fluctuations within each bubble universe transitioning out of inflation provide an infinity of differing initial conditions for the causally disconnected parts of each bubble universe that form a beyond-the-horizon multiverse. The inclusiveness of all the random possibilities in the bubble multiverse is reminiscent of the many worlds multiverse. In both cases, the countless random choices within the quantum possibilities insure that all possibilities are covered and that what *could* happen, *did* happen somewhere.

In fact, the sets of distinct universes in the many-worlds and bubble multiverses, respectively, are indistinguishable.[19] According to Tegmark, "...the [many worlds multiverse], if it exists, adds nothing new beyond the [bubble and beyond-the-horizon] multiverses—just more indistinguishable copies of the same universes, the same old storylines playing out again and again in other quantum branches."[20]

Perhaps all this is simply academic. Who cares what's out there if we can't ever possibly see it and test it? Or, to the extent these concepts are fascinating, perhaps they belong more in the realm of metaphysics than physics. American theoretical physicist Lee Smolin has gone so far as to claim that the beyond-the-horizon and bubble multiverses are concepts that, by construction, are not falsifiable and, therefore, *not scientific*.[21]

Smolin, who now works at the Perimeter Institute for Theoretical Physics in Waterloo, Ontario, is a bit of a rogue spirit known for questioning fashionable ideas in physics. Recently, he published a somewhat-controversial book critical of string theory.[22] Of course, rogue does not mean wrong, and hard questioning is an important part of the scientific process. (As an aside, Smolin has created an interesting multiverse concept that he maintains *does* satisfy rigorous constraints for a scientific theory, which I'll discuss in Chapter 8.)

Certainly, through our current understanding and using the definition of a universe as something potentially causally connected to us, we'll never witness regions outside our cosmic horizon—outside regions of space that can, in principle be causally connected to us. In that sense, there is no arguing against Smolin. However, it is common and productive scientific practice to use an experimentally supported framework of concepts to infer things that have not been observed directly. It is in this spirit that I call the beyond-the-horizon and bubble multiverses scientific concepts. Certainly, aspects of our modern view of the universe are empirically testable and falsifiable and the multiverse concepts follow from those things.[23]

Cosmologists such as Garriga and Vilenkin and Tegmark believe that if inflation happened, the beyond-the-horizon and bubble multiverses are unavoidable given our current understanding of physics.[24] Inflation implies they exist.

So, did inflation really happen? Recall that the case for inflation is compelling. The big bang is strongly supported by evidence such as the CMB and the abundance of light elements in the universe. Inflation fixes a few serious problems with the big bang model quite naturally, including the horizon problem and the flatness problem. Still, these elegant solutions to troubling aspects of the big bang only amount to indirect evidence for inflation.

The most direct and significant evidence pointing toward an inflationary beginning to our universe comes from the detailed study of fluctuations in the CMB. Recall that in the inflationary picture the slight temperature variations in the CMB arise from quantum fluctuations in energy that are stretched to a cosmic size by inflation. These fluctuations in energy eventually translate into variations in the matter density that we now see as the large-scale structure and distribution of matter in the observable universe. As we saw in Chapter 6, cosmologists quantify the fluctuations in the CMB by looking at the degree to which the CMB temperature changes for the different sized regions, which is what they refer to as a power spectrum. Theoretical models for the big bang that include inflation at the beginning happen to produce a power spectrum, a measure of how frequently the different size structures occur, for the structure in the CMB that is different from other models on the market.[25] This is something predictive that *can* be tested experimentally.

In 2003, the team analyzing the data from NASA's WMAP[26] satellite held a press conference where they announced that the power spectrum of fluctuations in the CMB agreed well with the inflationary model. Astrophysicist John Bahcall, world-renowned in part for his work on the theory of the nuclear processes in the sun, gave the concluding remarks at the press conference. Bahcall said, "WMAP has confirmed with exquisite precision the crazy and unlikely scenario that astronomers and physicists cooked up based upon incomplete evidence."[27] That unlikely scenario supported by the WMAP data and referred to by Bachall is the inflationary big bang model with dark energy and cold dark matter—the concordance model.

Perhaps this "unlikely scenario" is just crazy enough to be true! Or perhaps not. The concordance model is something of a patchwork quilt of ideas. There are other cosmological possibilities to consider— even ones that lead to a multiverse. These other possibilities are the subject of the next chapter.

8
Doing a Little Gravity

few weeks ago I was enjoying a beautiful end-of-summer day (yes, we *do* have beautiful days in Rochester) having coffee and conversation with a non-physics friend when one of my physics colleagues, Adrian Melissinos, happened by and started chatting. Adrian is one of my favorites among physicists. He hails from Greece and spent a stint in the Greek navy before entering grad school at MIT. Being the type of person who enjoys life and loves physics and learning about the universe, Adrian's enthusiasm is infectious. He has made a career in physics out of looking for things under rocks well off the beaten path. Fortunately, he is blessed with the creativity and drive to have made this work. I asked Adrian what he'd been doing over the summer. Adrian grinned and said, "Oh, this and that, you know. I've been doing a little gravity." And the conversation flowed from there.

It wasn't until Adrian walked away that my friend snickered and said, "*Doing* a little gravity? And just *how* do you *do* a little gravity? Do you have any idea how crazy you guys sound?"

Point taken. I guess it sounded a bit strange. Of course, I knew what Adrian meant by "doing a little gravity." Adrian works with a big collaboration on an experiment called the Laser Interferometer Gravitational Wave Observatory (LIGO), a large project designed to detect gravitational waves, which are extremely tiny distortions in space-time predicted by general relativity. Also, I think Adrian may be spending some time thinking about the search for extra dimensions at the LHC. "Doing a little gravity" means he was working on these things over the summer.

What's more interesting to consider is *why* Adrian might be doing a little gravity. I mean with general relativity, don't we know all about gravity already? Perhaps not. Gravitational waves are predicted by general relativity, but have yet to be detected. Detecting them will be a big step in physics, and it may lead to a new field where scientists study distant objects by their movements rather than by seeing them. This will be important for studying interesting regions of space that are obscured by gas clouds in space, such as the center of our own galaxy. As for extra dimensions, an area of physics known as string theory has opened up all sorts of new possibilities concerning the number of space-time dimensions and the behavior of gravity at very short distances. All this strangeness concerning gravity, the number of dimensions, and how space is warped has spawned several visions of the multiverse. That's where we are headed next.

For gravitational strangeness, it's hard to beat a black hole. Have you ever wondered what happens *inside* a black hole? According to one of my students, that's where the toes wiggle! After all a black hole is what you get in a black sock, right? Trust me. In a physics class, a student like that can help things flow along.

There are people—other than this particular student of mine—who have pondered the question. The answer is that we really don't know what happens inside a black hole. We can't see inside black holes. Even if we could see inside a black hole, quantum gravity likely plays an important role in what happens there, and we don't yet have a full understanding of quantum gravity. Still, that doesn't keep physicists from extrapolating the physics we *do* understand into that realm and speculating on what might happen there.

Recall from way back in Chapter 2 that general relativity views gravitation as the result of the curvature of space-time due to the presence of energy or mass. A black hole is formed when energy (or mass) is crammed into such a tiny volume that the gravitational field bends the surrounding space-time so much that that light does not escape. Approaching the black hole, the gravitational field strength, and the corresponding bending of the space-time, becomes stronger and stronger. There is a characteristic distance from the center of the black hole inside of which light cannot escape because gravity and the bending of space-time is too strong. This distance is known as the "event horizon." At the moment a black hole is formed, everything inside the event horizon becomes causally disconnected from the space outside the event horizon. Things can fall inside a black hole, but nothing can emerge from inside a black hole. In addition, according to general relativity, time is distorted by such a large gravitational field. To outside observers, a clock falling in a black hole would slow as it approaches the event horizon, and time would appear to stop at the event horizon. To observers holding the clock and falling into the black hole, things would seem to proceed normally and they would not notice when they passed through the event horizon.

Of course, this conceptual picture is based on thought experiments. We are playing pretend again. In truth, the gravitational field near a black hole is so strong that differences in the force across things—the so-called "tidal forces"—would rip whatever it is (observers, clocks, or anything else) to shreds. The observers holding the clock would be ripped to bits long before they could witness passage through the event horizon.

In principle, a black hole can be formed from things of any size as long as the energy/mass is compressed inside a small enough volume of space. Every time particle physicists turn on a new accelerator, a series of articles appears in the press (and the occasional lawsuit aimed at stopping the project) claiming that the accelerator will form black holes. Do a Web search for "LHC black hole" and see what happens.

In fact, it's not that easy to make a black hole. The energy density needed to create one is extremely high—much higher than that which is available at the LHC, or any other accelerator we are currently discussing. There is theoretical speculation that conventional gravity may give way to something stronger at very tiny distances, but if that were to happen the black hole would decay away quickly through quantum processes we understand. The fact that the moon exists is a strong indication that our accelerators will not create an Earth-eating black hole or a different sort of strange particle that might prove to be disastrous.[1] It's been around for some 4.6 billion years and under constant bombardment by energetic particles from astrophysical sources, some of which are far more energetic than we can create in our accelerators.[2]

The black holes in our universe are thought to be formed in the centers of galaxies and in the last stages of the life cycle of massive stars. In a nutshell, stars are formed from clouds of gas in space— mostly hydrogen—that condense under the attractive force of gravity. As the cloud condenses, it heats up. This heating is nothing special to stars; it happens when gas is compressed. Feel the nozzle next time you pump up a bicycle tire and you'll find it's hot. At any rate, the gravitation slowly pulls the gas inward, and things get more and more dense and hot at the center of the cloud. Eventually, it becomes hot enough that nuclear fusion reactions begin in the center of the new star, converting the hydrogen into helium and releasing a vast amount of energy in the process. This is the same process that powers hydrogen (thermonuclear) bombs.

The energy produced by fusion in the center of the star pushes outward through a phenomenon called "radiation pressure" and balances the gravitational pull inward. This equilibrium between the energy from hydrogen fusion pushing outward and gravity pulling inward can last for many billions of years, depending on the size of the star. Our own sun is less than halfway through this stage of its life, which is expected to last around 10 billion years.

Eventually the fuel at the center runs out, and the core of the star collapses, gets hotter, and begins to burn heavier elements in fusion reactions. I'll spare you the details. What happens in the final stages of a star's life cycle depends very much on the mass of the star.[3] In the end, if the star is massive enough, when things cool down in the center, gravity wins and the star collapses. As more and more material is compressed into a smaller and smaller volume, the gravitational field in the region close to the star becomes stronger and stronger. Eventually, it becomes strong enough to form a black hole.

At the moment a black hole is formed, the region inside the event horizon is causally disconnected from the region outside the event horizon. We can never look inside a black hole formed in our universe. That said, it is known what to expect in the context of classical general relativity. Inside the black hole, we would not perceive things to be strange, provided we could ignore the gravitational tidal forces ripping us to little bits. Time would not seem to stop for us. Inside the black hole, general relativity says that there is a singularity—a point where the curvature of space-time (that is, the gravitational force) is infinite. Particles sucked into the black hole have trajectories that take them into the singularity and that's that. There's no future beyond that for those particles.[4]

Or perhaps there's more to the story. As the energy density grows near the region of the singularity, there comes a point where the quantum wavelike character of typical photons (when they reach an energy of roughly 10^{28} eV) is on the same size scale as the local curvature of space-time caused by that photon (because it is a form of energy and energy bends space-time). Recall that this is known by physicists as the Planck scale. At this point, quantum effects may modify what is expected from general relativity, and it's not reasonable to extrapolate our thinking beyond that point without a theory of quantum gravity. It has been speculated that quantum gravity effects might eliminate the singularity and cause the collapse to reverse leading to expanding space-time, which may, according to Lee Smolin, "result [in] the creation of a new expanding region of space-time, which may grow and become, for all practical purposes, a new universe."[5]

Smolin has proposed a fascinating multiverse concept built on this strange idea of universes born inside black holes.[6] It is what I call the "fecund multiverse."

There are two postulates assumed by Smolin in this theory. The first is that inside black holes the effect of quantum gravity causes the collapse of the star to reverse, creating a new region of expanding space-time—a new universe—inside the black hole. As strange as it may seem, that new universe is totally isolated by the event horizon from the universe in which the black hole was formed. After all, black holes are outta sight, right? Sorry. Anyway, this universe-in-a-bottle idea is strange, but let's just go with it for now.

Smolin's second postulate is that the physical constants, or parameters, that characterize the universe are changed slightly during the quantum gravity space-time "bounce." When the collapse is reversed, the characteristics of the space-time that determine the strength of the various forces and the nature of the particles in that universe are shaken up a bit, but not too much. The universe born inside the black hole has physical parameters that are very similar, but not quite identical, to the universe in which the black hole was formed.

Smolin argues that the fecund multiverse exhibits cosmological natural selection. The trick to this is that a universe formed within a black hole might spawn black holes that give rise to universes within them, and so forth. According to Smolin:

> As our own universe contains an enormous number of black holes, there must be enormous numbers of these other universes. There are at least as many as there are black holes in our universe [perhaps as many as 10^{18}], but surely if we can believe this we must believe there are many more than that, for why should not each of these universes also have stars that collapse to black holes and thus spawn new universes?[7]

Natural selection comes in because universes with parameters optimal for the production of black holes will create more black holes that will all create universes with parameters close to optimal for black hole production. Universes where the physical parameters are such that it is difficult or impossible to produce black holes do not "reproduce"; no black holes are formed to create new universes with similar parameters.

In the fecund multiverse, quantum fluctuations, or whatever strangeness theoretical physicists cook up, give rise to all the possible universes with all the possible sets of parameters. Most of these do not exhibit inflation or possess parameters that lead to the physical properties that allow for the formation of stars and black holes. However, eventually a universe is formed where black holes can form. That leads to more universes (inside the black holes of the first) where black holes form, and within those are more universes with black holes, and on and on. Before long, the multiverse is populated with universes where the physical parameters are close to optimal for black hole production.

What does it mean to have parameters close to optimal for black hole production? Well, for a black hole to be produced, stars have to form and live out their life cycles. This requires a universe that does not expand too slowly or too fast, which means the cosmological constant is in a narrow range. The weak nuclear force has to be just the right strength. Neutrons have to be slightly higher in mass than protons.[8]

Smolin pushes the fecund multiverse as a testable, scientific multiverse. He argues that ideas like the bubble multiverse are not scientific because they are not falsifiable. The fecund universe, on the other hand, is falsifiable. If cosmological natural selection happens, then the typical universe in the fecund multiverse should be near optimal for black hole production. If that's the case, the physical parameters of our universe should be near optimal for black hole production, because the ideas of scientific naturalness and simplicity require our universe to be a typical member of the multiverse. According to Smolin, "The parameters of the standard model of elementary particle physics have the values we find them to because these make the production of black holes much more likely than most other choices."[9] If it is shown that our universe is very far from being an ideal place to create black holes, the fecund multiverse cannot reflect reality.

Is our universe an optimal factory for black holes? Smolin argues that it is an open question and that, to date, his idea of cosmological natural selection has not been falsified.[10] At least one renowned cosmologist and astronomer—Joseph Silk—disagrees.[11] Science is like that. Smolin's idea is falsifiable. In time, perhaps a consensus will emerge on whether or not what we see in our universe is consistent with the fecund multiverse.

Regardless of the degree to which our universe is optimal for black hole production, the details of the fecund multiverse theory are based in quantum gravity. General relativity works well at large distances, where space can be thought of as very smooth. However, at small distances things must change. Recall that empty space is a seething sea of virtual particles. At the smallest distance scale, those virtual particles can have enormous energies/masses and not run afoul of Heisenberg's uncertainty principle. Presumably, these virtual particles cause space to be anything but smooth at small distance scales. According to one of the greatest theoretical physicists of the 20th century, John Wheeler:

> What we think of as smooth, simple space, is really a "wiggly" business. I don't know any better image for it than the look of the ocean. As one comes down from a plane high above the ocean, that seems to be a perfectly smooth surface. You come down closer, you see the waves, and as you get still closer you see the waves breaking and you see foam. I think it must be the same in the geometry of space, for all our everyday experience, the geometry of space is smooth and flat. But as we examine it more closely, it must show oscillations. And still more closely, it must show foam, a foam-like structure.[12]

At these small distances, the equations of general relativity do not work and a new theory of quantum gravity is needed.[13]

Many physicists—though, perhaps not Smolin[14]—believe the key to quantum gravity is in the branch of physics known as "string theory." It happens that few things cause as much discussion in a physics department faculty meeting than a suggestion that the department should hire a string theorist. At one extreme are physicists who are of the opinion that string theory is the most promising approach to building the long-sought theory of everything, including quantum gravity, while those at the other extreme think string theory is little more than an abstract area of mathematics that has no place in a physics department. I can't get very excited about the argument. Regardless of whether or not physicists ever succeed in using the ideas of string theory to build a realistic theory of quantum gravity or a theory of everything, the area has spawned a great deal of intriguing work, including interesting ideas about gravity, dimensions, and the multiverse. There is value in that whether the people who do it are housed in a physics building or a mathematics building.

String theory began in the late 1960s with no pretense of being the framework for a theory of everything. In the words of one of the originators, Leonard Susskind, "Today's modern String Theory is all about the elusive unification of quantum mechanics and gravity, over which physicists banged their collective head for much for the twentieth century. ...it started out much more modestly as a theory of hadrons."[15]

Recall that hadrons are particles made up of quarks. In the 1960s, physicists were struggling to make sense of the zoo of particles uncovered by experiments at particle accelerators. Susskind stumbled on the idea of treating hadrons as if the quarks inside them were connected by something like a rubber band or a string, and found that he was able to understand some of the basic results of the day. Two other physicists worked on this idea independently around the same time: Yoichiro Nambu,[16] a Japanese-born American physicist working at the University of Chicago, and a Dane named Holger Bech Nielsen at the Niels Bohr Institute in Copenhagen. Together these three are considered the founders of string theory.

It turns out that physicists were not successful in using string theory to construct a model of hadrons. Instead, field theory succeeded in that with the theory of quantum chromodynamics, as we saw in Chapter 5. Still, one man's trash is another man's treasure. One of the "problems" with string theory as a theory of hadrons was the emergence of a particle with very particular characteristics that did not correspond to a known particle. Then, in 1974, physicists Joël Sherk of the Ecole Normale Supérieure in Paris and John Schwarz of the California Institute of Technology realized that this peculiar particle had exactly the characteristics of the quantum of the gravitational force, known as the graviton.[17] With this realization, the early string theorists began to work on using string theory as a basis for a theory of quantum gravity. About this, Leonard Susskind remarked, "If you can't make a String Theory of hadrons because the theory insists in behaving like gravity, then let gravity be described by String Theory."[18]

Before we move on, perhaps it's good to ask: What is string theory, and what does it mean for a particle to emerge from the theory? String theory is named aptly because it is built on the idea that fundamental particles are not point-like, but rather string-like objects. The idea is that different resonant vibrations of the string correspond to different particle states in nature. To see what I mean, imagine a string tied between two posts, like what you might find on a guitar. Such a guitar string plays certain discrete musical notes—and only those notes—clearly when plucked. Those notes correspond to vibrations like the ones pictured in Figure 8.1 on the following page. The notes played by the string can be varied by changing the thickness,[19] the length, and/or the tension of the string.

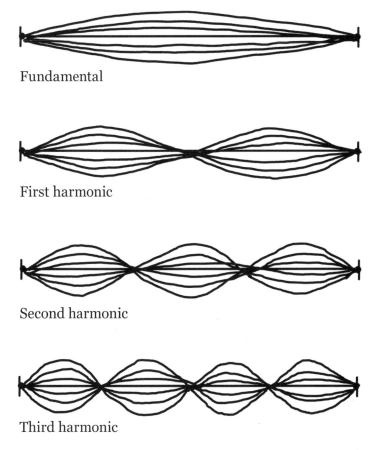

Fundamental

First harmonic

Second harmonic

Third harmonic

Figure 8.1: Illustrations of the most common string vibration patterns for a guitar string. These so-called resonant vibrations yield clear, clean notes. The "fundamental" mode of vibration corresponds to the lowest pitch sound the string can play. The first harmonic pattern shows what the string vibration pattern looks like for the next highest pitch sound played clearly by this string, the second harmonic pattern gives the next highest pitch sound above the first harmonic, and so on.

The strings used by physicists to model particles have some similarities to guitar strings. They are extended in length, and the string vibrates in only certain ways. Of course there are big differences between a string theorist's string and the ones used by Eric Clapton. The strings used in string theory are very tiny—only about 10^{-35}

meters in length—which is the distance where the quantum structure of space is thought to be important. The mathematical theory,
or framework, that describes the vibration of this little string is quite
different from the classical wave theory used to describe vibrations
on a guitar string, too. In string theory, the equations must play well
with relativity and quantum mechanics. This complicates things
considerably. Physicists meeting with research success in the area
of string theory typically have extraordinary mathematical talent.

It was noticed by physicists that the mathematics describing
quantized, relativistic excitations (physics lingo meaning vibrations)
was the same as the mathematics used by particle physicist to describe subatomic particles. In other words, the discrete types of vibrations occurring on these little strings could be thought of as particles. It was this correspondence that got physicists excited about
using strings to describe particles back in the late 1960s.

The world of strings is richer than you might first guess. One of
the nice things about working with mathematics rather than real
guitar strings is that you can imagine some possibilities that might
not be so practical in a guitar factory. Strings in string theory can
be open-ended or closed loops. They can have an extraordinarily
high tension. String theorist Brian Greene says, "According to string
theory, the properties of an elementary 'particle'—its mass and its
various force charges—are determined by the precise resonant pattern of vibration that its internal string executes."[20] That pattern is
determined, in part, by the properties of the string: open-ended versus closed loop, string tension, and so forth.

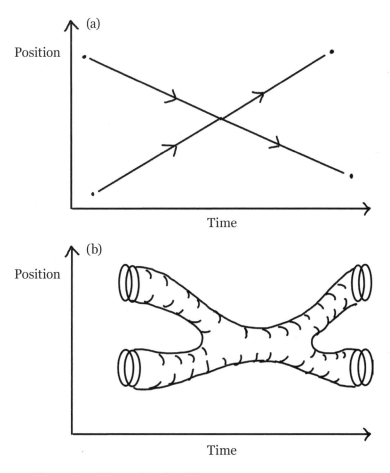

Figure 8.2: These sketches illustrate two particles interacting. The distance between the particles is plotted on the y-axis, and time is shown on the x-axis. In (a), two point particles interact and—no surprise— the interaction takes place at a point. In (b), two particles pictured as closed loops, or strings, interact. The paths taken by the particles are pictured tubes. In string theory, the interaction takes place over an extended region, as shown, rather than at a point.

The optimism that string theory might hold the key to quantum gravity is fueled in part by the fact that the graviton emerges naturally in the theory. It also happens that strings have a different short-distance behavior than the point-like particles in field theory. This is illustrated in Figure 8.2. In quantum field theory, particles interact at a point as in Figure 8.2(a). As the distance between the particles approaches zero, the mathematics that describes that interaction makes no sense for a quantum field theory including a graviton. To see why this might happen, let's consider a little example. Suppose the force between two particles drops as they grow further apart by 1/distance. In this case, for example, the force between the particles drops by a factor of 10 in strength when they are 10 times further apart. Conversely, as the two particles approach each other the force becomes very large. As the distance in the denominator shrinks to zero, as is the case with point-like interactions, the force becomes infinite. Things are a bit more complex in real life and theorists have developed ways of handling many types of things like this. Still, in the case of the graviton, the math goes wacko when the distance shrinks to zero.

In string theory the particles are extended in space because they are made of a string. Consequently, the interactions between the particles are spread out a bit in space as well, as illustrated in Figure 8.2(b), where the motion of the strings are represented by the tubes they form as they move. This leads to much better behavior of the mathematics, which in turn makes it possible, in principle, to construct a theory of quantum gravity using strings. According to Brian Greene:

> When the force involved in an interaction is the gravitational force...this complete packing of the force's punch into a single point leads to disastrous results, such as the infinite answers.... Strings, by contrast "smear" out the place where interactions occur.... This spreads out the force's punch and, in the case of the gravitational force, this smearing significantly dilutes its ultramicroscopic properties—so much so that calculations yield well-behaved finite answers in place of the previous infinities.[21]

Don't get the idea that switching from points to strings was some magical fix that led to a theory with no problems. In fact, string theorists have struggled with many tough issues. In addition to the difficulty of the calculations, the string theories of the 1970s predicted many particles not found in nature and were fraught with solutions that appeared to have negative probabilities, which makes no sense. Of this time, Lisa Randall, a well-known theoretical physicist at Harvard, says, "Working on string theory in the 1970s required individuals who were either very determined or somewhat crazy."[22]

Nevertheless, some hardy souls persevered. In the 1970s, a number of physicists (Pierre Ramon, John Schwarz, André Neveu, Ferdinando Gliozzi, Joël Sherk, and David Olive) contributed toward developing a string theory that contains both fermions and bosons.[23] Because we know that nature includes both fermions and bosons, this development was seen as essential if string theory is to be used to describe reality. More than that though, they found that string theory could be supersymmetric—which is to say it possesses that fermion-boson symmetry mentioned at the end of Chapter 5, which might be the answer to the hierarchy problem in particle physics. String theories constructed in this way are called "superstring" theories.

In another development of the 1970s, John Schwarz and a British physicist named Michael Green found that a class of unphysical solutions plaguing superstring theory (nonsense solutions to the equations that correspond to particles with a negative probability) disappeared if the theory was cast in 10 dimensions: nine spatial dimensions and one time dimension. Then, in 1984, this same pair of physicists showed that these superstring theories in 10 dimensions could be formulated to avoid a particular form of a deep and serious technical issue that threatened to limit the usefulness of string theory.[24] Suddenly, the promise of string theory seemed to outweigh the difficulties. The framework of string theory was perceived to have within it the potential to be a theory of quantum gravity, as well as the basis of a theory for the other forces of nature that might provide conveniently the key to understanding what lies beneath the Higgs mechanism. In other words, physicists were hopeful that superstring theory might become a theory of everything—the ultimate goal in physics. In fact, so many physicists turned their attention to super-strings during the period between 1984 and 1986 that the time is often referred to as the first superstring revolution.

Revolution or not, 10 dimensions is hard to swallow. The last time I looked we live in three spatial dimensions plus time. That's only four dimensions, six short of what our string theory friends seem to think we should have. Clearly this is a problem if the theory is meant to describe reality. Fortunately, mathematicians and theoretical physicists don't feel so constrained by reality, or at least they're willing to play pretend with ideas to see what happens. It is quite possible to construct theories with more than four dimensions. The tricky part is to make all the extra dimensions imperceptible.

How in the world can you hide a dimension? Physicists call the process "compactification." To get an idea of how compactification works, let's consider a simple example of a two-dimensional world as represented by the flat sheet illustrated in 8.3(a) on the following page. Suppose we curl the sheet into a tube by attaching one edge of the sheet to the opposite edge as seen in Figure 8.3(b). The tube has a size dictated by its radius, which is labeled "r" in the sketch. If we view the tube from a distance that is very much larger than r, the tube will appear to be a one-dimensional object: a line instead of a sheet. We've managed to hide the second dimension. It's still there, but it cannot be perceived unless we look very, very closely.

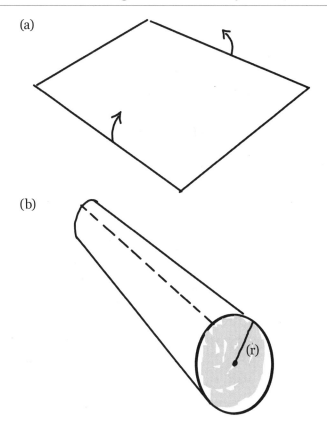

Figure 8.3: Consider the two-dimensional surface shown in (a). Suppose the two edges indicated in (a) are curled into a cylinder of radius r as shown in (b). If r is very small or if this cylinder is viewed from a large distance, it will look like a one-dimensional line. From a distance, the initial two-dimensional object will appear to look one-dimensional. That second dimension is said to be compactified.

The idea with compactification is to curl up the extra dimensions and shrink the "radii" to the point that we cannot perceive them—at least not with any experiments yet done. If the universe really works this way and we manage to probe to smaller and smaller distance scales, eventually we will uncover evidence for extra dimensions. This is something physicists at the LHC are actively investigating, because those experiments are sensitive to the smallest distance scales yet seen.

As strange as the idea of compactification of extra dimensions may seem, it isn't all that new in the world of physics. Early in the 1919, a Geman named Theodor Kaluza proposed to unify Einstein's general theory of relativity with electromagnetism through the introduction of a fifth dimension.[25] Interestingly, Albert Einstein refereed the paper (meaning he was asked by the journal to evaluate the paper for scientific merit) and delayed the publication for two years with his concerns about this extra dimension.[26] In 1926, a Swede named Oskar Klein proposed a compactified form for this extra dimension not so different in form from the compactification scheme illustrated in Figure 8.3.[27] In Klein's picture, every point in regular, extended three-dimensional space also has a small curled-up dimension that is too small to detect.

Unfortunately, the Kaluza-Klein compactification scheme was found to be inappropriate for string theory because once the six unwanted dimensions were curled up, the theory lost some of the critical technical features necessary for it to explain the details of the Standard Model.[28] In what turned out to be a huge breakthrough for string theory, a different and suitable compactification scheme was put forth in 1985 by an Englishman named Philip Candelas and three Americans, Gary Horowitz, Andy Strominger, and Edward Witten.

In the early 1990s, five superstring theories were known that had potential to be developed into a theory of everything. In 1995, one of the most influential theoretical physicists of our time, Edward Witten, showed that these five string theories were, in fact, deeply related and were likely different faces of a single theory that he dubbed "M-theory." Also in the 1990s, a number of physicists, most notably Joseph Polchinski at UC Santa Barbara, established that string theory means more than strings.[29] In fact, the theory can have structures of different dimensionalities, such as points, sheets, and three-dimensional spaces, within it in addition to strings. These structures are called "branes," which is a play on the word *membrane.* Strings can be confined to these branes, which are themselves embedded in higher dimensional space-time.[30] These ideas of branes and M-theory launched a flurry of activity that is now called the second superstring revolution.

Where are we after this brief, and all-too-encapsulated, history of string theory?[31] And what do all these strings and brane-twisting concepts (sorry, I couldn't resist) have to do with the multiverse? In string theory, the particles and forces and physical constants of nature are determined by the details of the compactification of the extra dimensions. According to string theorists Raphael Bousso and Joseph Polchinski, "In both the Kaluza-Klein conjecture and string theory, the laws of physics that we see are controlled by the shape and size of additional microscopic dimensions."[32] It was a dream of many physicists that the details of M-theory might be worked out and provide us with the long-sought theory of everything. Further, this theory of everything—so people hoped—would have a most natural and obvious compactification solution, which would lead directly to the Standard Model and the constants of nature we see in our universe.

Is this idea crazy? Not really. As we saw in Chapter 7, it is natural for any system to settle into a state that minimizes the overall energy. Physics is full of things like this. Springs don't stay sprung unless they are held in place by something else; water flows downhill; and like-sign electric charges are repelled apart.

Unfortunately, there are a vast number of ways in which the structures of the string theory can be put together and compactified. With respect to this, Bousso and Polchinski state:

Yet the vast collection of solutions are not all equal: each configuration has a potential energy, contributed to by the fluxes [the way the forces flow in the space-time], branes and the curvature itself of the curled-up dimensions. This energy is called the vacuum energy, because it is the energy of the spacetime when the large four dimensions are completely devoid of matter or fields. The geometry of the small dimensions will try to adjust to minimize this energy, just as a ball placed on a slope will start to roll downhill to a lower position.[33]

In case you're wondering, this vacuum energy is the same sort of thing we discussed in Chapter 7. There are configurations of the string vacuum that can cause inflation to take place. There are configurations of the string vacuum where inflation does not take place. The physical constants and the types of particles and forces in the universe depend on the details of the string vacuum—how the extra dimensions are compactified and what structures are present.

A physicist thinks about all these different configurations of the string vacuum in terms of parameters that specify the geometry. For example, in Figure 8.3, the radius of the curled-up dimension is a parameter—a choice—that determines how tightly that dimension is curled. The size of that radius—the degree or tightness of that curl—in turn, helps to determine the characteristics of our physical universe and the total energy stored in that configuration of space-time in the vacuum.

To get a handle on the energy stored in the vacuum, let's consider something much more straightforward. Imagine a ball rolling down a hill into a small valley. When the ball is high on the hill, it possesses what a physicist calls gravitational potential energy. This means that, by virtue of its height, it can roll down and convert that gravitational energy into energy of motion. If the ball is sitting still at the bottom of the hill, its gravitational potential energy is zero, or, a physicist would say, it is minimized. In this example, the energy of the ball depends on its location in the valley. If you were looking at a map of the valley, that location would be specified by two coordinates—say, where the ball is along the north-south direction and where it is along the east-west direction. Those two coordinates are parameters. If you specify those two parameters, the location of the ball is specified and that, in turn, determines its height and energy given the physical layout of the valley.

If you think about this example in terms of what Bousso and Polchinski tell us, a ball starting up high on the valley wall will naturally roll down until it is at the low point of the valley. Or will it? Suppose up on the hill overlooking the valley, there's a flat area with a dip. If the ball begins its travels near that dip, it might roll into the bottom of the dip and stop there instead of rolling down to the bottom of the hill. A physicist would call the bottom of that dip a local minimum in the potential energy of the ball and, barring some other event happening, the ball could very well be stuck in that dip forever.

The string theory vacuum is vastly more complex than the ball and valley example. There are a great many parameters associated with the compactification of the extra dimensions. Imagine mapping out all the possibilities for how the extra dimensions might be arranged during compactification, and tracking the energy of each configuration and how it depends on the numerous parameters specifying that configuration. With that information, you can make a big, multi-dimensional map of how the energy stored in the vacuum varies with the parameters. In reality, with the large number of parameters to vary, it is not possible to visualize this map in its full glory.

However, if you fix all but two of the parameters in this complex space of string theory possibilities and examine how the energy changes as the two remaining free parameters vary, then you'll end up with something like our map of the valley. One direction along the map—say, the east-west direction—represents one of the varying parameters while the other direction—north-south—represents how the other free parameter varies. Each point on the map represents a position where each of the two varying parameters is fixed to a specific value, sort of like latitude and longitude on a real map of the Earth. Imagine that at every point on the map a number is given that represents the energy of the string theory vacuum, much in the same way the height is shown in a topographical map. Even in this simplified case, the topographical structure on the map is likely to be much more complex than a simple bowl-shaped valley. It will look more like a topographical map of a mountain range, with countless peaks, valleys, ridges, and canyons in the energy. The map is full of local minima where the "ball might get stuck." Each of these local minima—a local low spot in the energy density of vacuum—represents a different geometrical configuration of the string theory vacuum that might exist in some stable or quasi-stable form.

Remember that we elected to limit the problem for simplicity. The real map has many more possibilities. In fact, string theorists estimate that there might be as many as 10^{500} possible configurations of the string theory vacuum.[34] Leonard Susskind calls this huge set of possibilities the "string theory landscape"[35] or the "cosmic landscape."[36] According to Bousso and Polchinski, "...the laws of physics that we see operating in the world depend on how extra dimensions of space are curled up into a tiny bundle. A map of all possible configurations of the extra dimensions produces a 'landscape' wherein each valley corresponds to a stable set of laws."[37]

How does this string theory landscape relate to the multiverse? It fits in quite elegantly with the bubble multiverse. Recall that, in the bubble multiverse, the greater reality consists of an eternally inflating matrix populated by ever-receding regions where inflation stops or slows due to quantum fluctuations in the inflaton field. In these regions where the inflation changes character, it is possible that something like the big bang happens and a universe more or less like ours might be born. When that happens the character of that local space-time, à la string theory compactification, will determine the nature of the dimensionality and particles and forces of that universe. It might be a very different universe from that which we know and love in our universe. About this, Max Tegmark says:

Although the fundamental equations of physics are the same throughout the Level II [bubble] multiverse, the approximate effective equations governing the low-energy world that we observe will differ. For instance, moving from a three dimensional to a four-dimensional (non-compactified) space changes the observed gravitational force equation from an inverse square law to an inverse cube law. Likewise, breaking the underlying symmetries of particle physics differently will change the lineup of elementary particles and the effective equations that describe them.[38]

To believers in the landscape, anything that can happen, does happen. There's no particularly special grand solution in a theory of everything that explains how and why our universe is exactly the way it is. To the landscapers, our universe is the way it is because all possibilities happen, and this is what happened here. There's no reason to continue to pursue that long-sought goal of the unique theory of everything.

To other physicists, the landscape of string theory is an abomination. Burt Richter, one of the great experimental particle physicists of our age, says, "Susskind and the Landscape school have given up. To them the reductionist voyage that has taken physics so far has come to an end. Since that is what they believe, I can't understand why they don't take up something else—macramé, for example."[39]

Landscape or macramé? We'll come back for more on this dispute and some of the philosophical and religious implications of the landscape in Chapter 9.

Beyond the landscape, string theory has inspired a very different and intriguing vision of the multiverse. This one is something of a rebirth of the oscillating big bang multiverse discussed in Chapter 6, but it is built using the brane structures of string theory invented in the 1990s.

This theory, called the "ekpyrotic universe," had its inception at a seminar given in August 1999 by a theorist named Burt Ovrut at a workshop in Cambridge, England.[40] The topic of Ovrut's talk was a bit of string theory weirdness dreamed up by Ed Witten and Petr Hořava at Princeton. In the audience of the talk were a couple other physicists whose interest was piqued, and, as happens sometimes at scientific gatherings, the ideas started flowing.

The concept explored by Ovrut in his seminar was that of three-dimensional branes wafting around in a higher dimensional space called the bulk. Recall that a brane is a structure in string theory with some dimensionality that can be embedded in higher dimensions. Imagine, for example, a two-dimensional sheet existing in three dimensions—like the paper before you. You can have multiple sheets that are all separated by the third dimension. If you were a little beast that lived on the surface of a sheet of paper, you'd never see those other pages, separated as they are from your world by that other dimension.

In the seminar, Ovrut discussed three-dimensional branes—sometimes called "braneworlds"—separated by one or more of the invisible dimensions of string theory. It is possible to construct models where some types of particles and forces are trapped on the brane, while other types of particles or forces might not be so constrained. So, for example, you can imagine a three-dimensional world—a brane—similar to the one in which we seem to live, where the particles and forces with which we are familiar are all trapped with us in the brane. This braneworld might be separated by a very tiny amount from other braneworlds by one of the extra dimensions of the string theory. Harvard string theorist Lisa Randall describes the idea in her book, *Warped Passages*:

> Even if the universe does have many dimensions, if the particles and forces with which we are familiar are trapped on a brane that extends in three dimensions, they would still behave as if they lived in only three. Particles confined to branes would travel only along the branes. And if light were also stuck to the brane, light rays would spread out only along the brane. In a three-dimensional brane, light would behave exactly as it would in a truly three-dimensional universe.[41]

Not all forces can be confined to a brane. Gravity is an integral part of space-time and must permeate all the dimensions if general relativity is correct.[42] Also, within these braneworld models, the fact that gravity bleeds out into the other dimensions is a way of explaining the weakness of gravity as compared to the other forces of nature.

Sound crazy? Perhaps it is. Lawrence Krauss, one of the more dynamic and public faces of physics today, says about braneworlds, "In some sense it is appropriate that this research area does sound like science fiction, because most of it probably is. What is too often underappreciated about science is that almost all of the ideas it proposes turn out to be wrong. If they weren't, the line between science and science fiction would be much less firm."[43]

The picture of multiple braneworlds wafting around in the bulk *almost* qualifies as a multiverse by our definition. It only fails because the different branes can interact with each other via gravitation. Gravity is a relatively weak interaction, though, and this set-up is still referred to by some physicists as a multiverse.[44] Still, this isn't the ekpyrotic universe promised earlier.

Back in 1999, it happened that in the audience of string theorists and cosmologists attending Ovrut's seminar in Cambridge were physicists named Paul Steinhardt and Neil Turok. Steinhardt was an old colleague of Ovrut's who had moved to Princeton, and Turok was a professor at Cambridge University. After the talk, both Steinhardt and Turok approached Ovrut with the thought that if the braneworlds could move around in the extra dimensions, they might collide, and, if they did collide, there might be a massive release of energy. Could it be that the big bang is nothing more than a collision between braneworlds?[45]

After the exchange in Cambridge, Steinhardt, Turok, Ovrut, and a graduate student of Steinhardt's named Justin Khoury spent the next couple of years looking at the cosmological implications of colliding branes. After much work, they came up with a fascinating new cyclic cosmology, which they eventually dubbed the ekpyrotic universe.

This new cyclic cosmology consists of two three-dimensional braneworlds separated by, but free to move in, one of the extra dimensions of the string theory. The separation between the brane-worlds is along one of the compactified dimensions and very tiny—perhaps something like 10^{-32} meters—because the dimension in which it happens is compact and unobservable. Because braneworld collisions are an important aspect of this cosmology, a spring-like force—repulsive when the branes are close and attractive when they are well separated—is hypothesized to exist between the braneworlds. The detailed nature of that force is critical to the theorists, of course, but it isn't all that important for our story. To see how this multiverse concept works, let's follow the evolution of one of the braneworlds through one cycle.[46]

There's a natural place to begin the cycle because this cosmology includes a stage where the braneworld would appear to be exactly like what we see today. This is no accident, of course, because the authors of this concept sought to build a realistic cosmology. Look around you. That's our starting point—*and* our ending point, because this is a cyclic model. At present, the universe is filled with clusters of matter and the space is expanding at an ever-accelerating rate. At this stage in the cycle, the braneworlds are slowly moving closer to each other, driven by the interbrane spring-like force acting between them. The potential energy stored in that interbrane force acts like dark energy causing the accelerating expansion of space in the branes.

Throughout a long period of time—say, a trillion years—the expansion of the space causes the matter to spread far and wide. The space in the brane becomes be very flat. In fact, any wrinkles in the space of the brane that were there due to events from earlier stages in the cycle are flattened out. As the branes approach each other, the potential energy stored in the interbrane force is reduced as if it is a relaxing spring. This causes expansion of space in each brane to slow but not stop.

In the moments before the branes collide, the small quantum ripples in space-time of each brane are amplified by the interbrane force. The branes become wrinkled by this process. Then the branes collide finally. With the ripples in each brane, the collision happens in some places before others. Some of the kinetic energy of the collision is converted into radiation within the branes. Throughout the brane there is energy released during the collision, though the temperature of the collision is uneven due to the uneven timing of the collision from spot to spot.

The branes rebound from the collision and move apart. After the collision the uneven temperature distribution within each brane looks look just like what we would expect from the big bang and is consistent with modern measurements of the CMB.

The branes reach a maximum separation one microsecond after the collision and then very slowly begin to move back together while the space within each brane expands. Big bang nucleosynthesis and star and galaxy formation proceed within each brane in the same way that it happens in the big bang model. Some 13.7 billion years after the collision we are back to where we started: a universe that looks like what we see now, and the cycle repeats.

Within each cycle, the expansion of the space in the brane is sufficient to spread the matter out so much that the matter density drops to zero. This eliminates the problem with entropy plaguing early cyclic big bang–big crunch models as discussed in Chapter 6. This route around the entropy problem is very similar to that used by Baum and Frampton in the cyclic patch multiverse, also discussed in Chapter 6.

The ekpyrotic universe is at least two multiverses in one. First, what takes the place of the big bang is a collision that happens over the entire volume of the three-dimensional branes at once, assuming we gloss over the tiny differences in timing due to the ripples. That volume is much larger than the observable universe. So, the ekpyrotic universe is a beyond-the-horizon multiverse, or what Tegmark would call a level I multiverse. Also, the ekpyrotic universe is cyclic. So, it is a multiverse in time. The "ekpyrotic beyond-the-horizon multiverse" gets reset every cycle leading to a vision of a reality that stretches forever into the past and forever into the future, at least to the extent that it is possible to imagine time as something outside of the context of what is happening in the colliding branes.

The ekpyrotic model is quite different from the inflationary big bang model. Steinhardt and Turok view this as a strength. The standard model of cosmology—the concordance model consisting of inflation, the hot big bang, dark matter, and dark energy—strikes some physicists as being cobbled out of random bits. Well-known astrophysicist John Bahcall called this scenario "crazy and unlikely" at the 2003 release of the WMAP results on the CMB.[47] Steinhardt and Turok say this is because "...the [concordance model] is a patchwork quilt sewn together from disparate ideas added over the previous two decades, plus the assumption of a particular, odd mixture of ordinary matter, dark matter, and dark energy."[48] About dark energy, Steinhardt says, "We can fit it into the [concordance] model, but we don't know what this so-called dark energy is. The standard model is definitely becoming more encumbered with time. It may still be valid, but the fact that we have to keep adding things is a bad sign."[49] The ekpyrotic universe does not involve inflation, and there is a natural explanation for the source of dark energy that drives the expansion of space. Also, it is a cosmology that fits within the string theory world.

Not everyone is convinced that the ekpyrotic universe is the right path. Alan Guth, the MIT professor who invented inflation, says, "I don't think Paul and Neil come close to proving their case. But their ideas are certainly worth looking at."[50] The scientific process is alive and well in this area of study!

To a non—string theorist mortal soul like me, the big question that comes to mind is: Why call this model the "ekpyrotic universe"? According to Steinhardt and Turok:

Paul approached classics scholars Joshua Katz from Princeton and Katharina Volk from Columbia University for advice. According to them, the new cosmological scenario sounded like the ancient Greek notion of ekpyrosis, in which the universe is born from out of fire. The word doesn't exactly roll off the tongue. Ovrut thought it sounded like some sort of skin disease. But we eventually settled for ekpyrotic universe, and the name has stuck.[51]

From doing a little gravity to the cosmic landscape, we've dodged some important questions. In the next stage of our exploration of visions of the multiverse, we'll take a closer look at a few of the deep philosophical, religious, and scientific questions that naturally arise when pondering the multiverse.

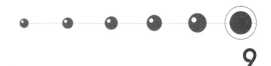

9

Copernicus on Steroids

f you ever happen to be walking the corridors of a university physics department late at night and you hear boisterous laughter, you may be encountering a graduate teaching assistant grading party. When exams are given to the big, introductory physics classes, it usually falls on a group of graduate teaching assistants to grade the exams. Often these exams require detailed mathematical solutions that must be inspected carefully by the graders. Small mistakes are so common that it is important to look at more than the final answer. Some mistakes indicate a far shallower understanding of the physics concepts than others, so a good grader must look at the problem and seek to understand what the student was thinking, so that he can award an appropriate amount of partial credit even if the final answer is incorrect. As they slog through stacks of papers, it's common for the graduate teaching assistants to share the more amusing answers with the other teaching assistants. Thus the laughter. It's not meant to be cruel or mean. It's done in an anonymous way, and it passes the time. Besides, laughing beats crying.

My own days as a graduate teaching assistant at Columbia are in the distant past now. Still, I remember grading one exam that has stuck with me. The details are foggy with years, but I vaguely recall the problem asked the students to calculate the mass of a baseball after being given some detailed information about the ball's motion. As I graded the exam of one student, I realized the mass they calculated for the baseball was something larger than the mass of the planet Jupiter. I'm sure I made some comment about this to the amusement of my grad student friends sitting at adjoining desks buried under their own piles of ungraded exams.

I think this particular exam answer has stuck with me through the years because it is a classic example of a student assuming the inviolability of the computer or, in this case, the calculator. Computers are stupid! They only calculate what you tell them to calculate and exactly as you tell them to calculate it. Where computers are concerned "garbage in gives garbage out." It's always critical to think about the results and ponder whether or not they make sense. If the answer doesn't make sense, it usually means there's a mistake in the calculation or an error in your thinking. It would take an extremely perverse physics professor to give an exam problem about baseballs where the mass of the baseball is set up to be bigger than the mass of Jupiter!

This is not a small issue in my life as a research physicist. We are constantly analyzing data using computers or simulating physics processes on computers. In the old days, when computers were less powerful and more of a novelty, we had to write most of the programs we used. We were forced to put an enormous effort into making sure we believed the results handed to us by computers. In those days, the burden was on the student to prove the program did what it was supposed to do. Somewhere in the 1990s things changed, at least in the graduate students I encountered working on particle physics experiments. By that time, big, self-contained software packages had evolved that, once customized, were able to run many of the complex calculations necessary for our experiments. With this development, it happened that many students became users of software they did not write or understand fully. Suddenly I had students doing complex analyses of data with these programs and coming into meetings with results that they hadn't thought about. They made claims with the attitude that if the computer says it's true, then it's true. This is something I've been fighting ever since. It's important to show many new graduate students that computers can't be trusted to think. Rather, the user must design test cases and check against known facts in order to build confidence in the calculations.

Are computers always going to be stupid? Perhaps not. I have seen modern computer algorithms that have been designed to allow computers to make reasonable decisions, corrections, and adjustments based on what they see for very particular and limited problems. In fact, a subfield of computer science deals with the creation of artificial intelligence. As computers become more powerful and the algorithms more advanced, and computer scientists make another breakthrough or two, artificial intelligence may come about.

Or perhaps artificial intelligence is *already here* and it is you and me. Could it be that I am a simulation that wrote this book and you are a simulated person reading it? No, this isn't a science fiction movie. It's an idea that I call the "computational multiverse." Here's the way it works. Given the incredible advances in computing power and algorithms in the last half century, it seems very reasonable to assume that we—or some other intelligent civilization—will have massive amounts of computing power to use for intriguing purposes. Just like humans now use computing power to create simulated worlds such as The Sims[1] or Second Life,[2] you can imagine that a being with massive computing resources might simulate a universe and that universe might contain conscious, simulated beings that can communicate and might have their own desire to simulate a universe. Before long, the conscious entities in the simulations—let's call them "Sims" following the Brown University philosopher, Brian Weatherspoon[3]—will vastly outnumber real conscious beings. Also, as in the video games of today, the simulations could have different physical laws and constraints. In fact, part of the fun for the creator of the simulated world might be the exploration of the outcomes of changes in the physical laws of the simulation.

The idea that our reality might be a simulation has been bounced around in the world of science fiction for some time. Perhaps the most famous example of this is the 1999 blockbuster movie *The Matrix*.[4] Recently, some serious thinkers have pondered this strange idea. Nick Bostrom, a Swedish philosopher who is the director of the

Future of Humanity Institute at the University of Oxford, published a paper in 2003 that argues one of the following must be true:

1. Humanity is likely to become extinct before reaching the level of technological advancement that we can simulate universes with Sims.

2. There is a fundamental limitation in computing that keeps advancing civilizations from simulating a universe with Sims or, for some unknown reason, a civilization that develops the capability to simulate a universe with Sims chooses not to do so.

3. "[W]e are almost certainly living in a computer simulation."[5]

Item 1 is really a statement that advanced civilizations become extinct before reaching the stage in technology where they can create simulations with the degree of sophistication we are discussing. That item isn't much fun to consider. Item 2 allows for there to be fundamental physical limitations in computing power that we've not yet encountered. That could be the case. We won't know until we encounter it. If 1 and 2 are not true, it isn't obvious why 3 must be true. The argument given by Bostrom is that once simulations have achieved the power to create Sims and simulated universes, those Sims will do the same and Sims within those simulations will do the same, and so forth. John Barrow, a cosmologist at Cambridge, summarizes the idea nicely in a recent paper: "Once this capability to simulate universes is achieved, fake universes will proliferate and will soon greatly outnumber the real ones."[6] This means that it is much more likely that we are part of a simulation than not.

Is there a way to tell whether or not we are part of a simulation? John Barrow speculates that, though the simulator's knowledge of the physical laws must be very great, it is likely to be imperfect.[7] This means the simulation is likely to develop drifting constants of nature or sudden glitches glitches glitches glitches glitches glitches glitches glitches glitches glitches. (Sorry, I couldn't resist.)

Bostrom's argument about the computational multiverse is intriguing. Still, it is a logical and philosophical argument, and, as far as I know, there's no evidence that we live in a simulated universe. Perhaps it is possible to do a scientific study searching for evidence that we are Sims by looking for clues such as those suggested by Barrow. For now, I've categorized the computational multiverse as a faith-based, non-scientific multiverse.

One of the interesting things about the computational multiverse is that the great programmers can dial up any sort of universe they wish. They can choose the physical laws of a simulated universe to have a very particular form and the physical constants to be highly tuned to particular values. Or not. All things are possible in simulations.

Of course the idea of a great programmer—or possibly I should say Great Programmer—is not new. The computational multiverse has nothing to do with religion. Still, it bears some similarity to the religious view of creation and control. Paul Davies, well-known theoretical physicist at Arizona State University, says, "...the denizens of a simulated virtual world stand in the same ontological relationship to the intelligent system that designed and created their world as human beings stand in relation to the traditional designer/creator Deity...but with God now in the guise—not of the Grand Architect—but of a Grand Software Engineer."[8]

How *do* God and the multiverse fit together? That the universe was created by God is, of course, a very strongly held opinion by many people. It is a matter of religious faith, and that is enough for many believers.

There have long been proponents of planes of existence that transcend the normal universe, particularly surrounding concepts of the afterlife. Many religions come equipped with planes that are heavenly or positive, and those that are hellish or negative. According to these beliefs, the disembodied essence or soul of a person exists in one or more of these planes after death, often with conduct during life determining the plane(s) of existence afterward. The ancient Egyptians had a complex vision where successful passage to one's reward in the afterlife required a mummified body, a sin-free heart, and knowledge of the Book of the Dead. Christianity teaches that heaven is a place of eternal life shared by the elect, or those receiving God's blessing, or those who have suffered sufficiently to be cleansed of sin; the many Christian sects disagree on exactly what it takes to gain this eternal life. Hindu cosmology says there are six heavenly planes, seven nether planes, and 28 hellish planes where souls dwell—the choice of which depends on the good and bad activities performed on the earthly plane—before being returned to the earthly plane in the form of a human or animal. In Islam, heaven is a happy place where immortal inhabitants enjoy the company of loved ones and many physical luxuries and pleasures. The ancient Greeks maintained there were 12 powerful gods living at the top of Mount Olympus and other gods living in the underworld called Hades and elsewhere. These gods acted like humans in many ways and shared many human vices. Visions of planes where gods and/or souls dwell that are outside the normal earthly universe have always been with us and are likely to always be with us. I call these ideas collectively the "multiverse of faith." You can choose to believe in them or not. I consider this a non-scientific multiverse because, to my knowledge, there is no scientific evidence to support the existence of these separate planes of reality.

We can't stop there on God and the multiverse. The relationship between religion and the multiverse is more complex than what we see in the multiverse of faith. The two get tangled together in the multiverse concepts inferred from science as well.

For those who look for evidence to support their faith, many feel there is little need to look beyond the world around us. They argue that the universe and life are far too complicated and interconnected to have arisen by natural means; the incredible complexity of the natural world must have arisen from the design and handiwork of an intelligent agent, usually meaning the Christian God, though that does not follow from the design argument alone. Proponents of this line of reasoning, called "intelligent design," argue that the observation of such inexplicable complexity constitutes scientific evidence for a supernatural origin. Most of these claims have focused on biological systems.[9] However, there is a variation of the intelligent design argument that looks for evidence in the laws and constants of physics. In the words of physicist and philosopher, Victor Stenger:

> The claim of evidence for divine cosmic plan is based on the observation that earthly life is so sensitive to the values of the fundamental physical constants and properties of its environment that even the tiniest changes to any of these would mean that life, as we see it around us, would not exist. The universe is then said to be exquisitely fine-tuned—delicately balanced for the production of life. As the argument goes, the chance that any initially random set of constants would correspond to the set of values that we find in our universe is very small and the universe is exceedingly unlikely to be the result of mindless chance. Rather, an intelligent, purposeful, and indeed caring personal Creator must have made things the way they are.[10]

This is an important discussion in the United States. If it is accepted that intelligent design has scientific validity, theories of supernatural origin must be taught in public schools along with the accepted scientific ideas such as evolution.

Is our universe fine-tuned to support life as we know it? Certainly it is true that, were particular physical constants or physical laws a bit different, it would have led to a universe where life as we know it could not have evolved.[11] According to Max Tegmark, "...most if not all the parameters affecting low-energy physics [typical everyday phenomena] appear fine-tuned at some level, in the sense that changing them by modest amounts results in a qualitatively different universe."[12]

The fusion processes and element synthesis in stars depends sensitively on the amount of hydrogen, deuterium, and helium formed in the early universe during big bang nucleosynthesis. If the weak force were slightly stronger, all of the neutrons in the early universe would have decayed into protons and the matter in the early universe would have been all hydrogen. If the weak force were slightly weaker, few neutrons would have decayed in the early universe and the matter in the early universe would have been all helium. Either way, the chain of fusion processes happening inside stars would have been broken, and the carbon in our universe would not have been formed.

If gravity were much stronger than it is, stars would burn faster. A shorter stellar lifetime leads to less time for complex life to arise from organic evolution.

If the cosmological constant were much larger than it is, the universe would have expanded too fast, and stars and galaxies would not have formed. If the cosmological constant were very negative, the universe would have collapsed long before stellar and biological evolution could have led to life as we know it.

Even the most ardent supporter of pure scientific methodology has to admit that some degree of bias in the physical constants and laws of physics must be present in our universe. As Stephen Weinberg points out, "Any scientists who study nature must live in a part of the landscape where physical parameters take values suitable for the appearance of life and its evolution into scientists."[13] We *are* here, after all.

Weinberg's words are a form of the anthropic principle first put forth by Robert Dicke in 1961, and reformulated and increased in scope by a number of authors since.[14] The anthropic principle comes in weak and strong varieties. The weak anthropic principle states, "The observed values of all physical and cosmological quantities are not equally probable but take on values restricted by the requirement that there exist sites where carbon-based life can evolve and by the requirement that the Universe be old enough for it to have already done so."[15] This form is not contentious and more or less states the obvious.

The strong anthropic principle, on the other hand, is quite contentious and is consistent with the idea of intelligent design. The strong anthropic principle says that "the characteristics of our universe are chosen to ensure the appearance of life and observers."[16] This, of course, immediately brings to mind the questions as to who or what made the choices and how they made the choices. These are questions that are considered well outside the domain of science in the eyes of most scientists.

The different forms of the anthropic principle have led to some confusion. The weak form is effectively what a scientist would call a selection bias. This means you only see what you are able to see with any experiment due to the nature and design of the experiment. A careful scientist has to minimize and correct for this bias in a good experiment. The strong form of the anthropic principle is (strongly) considered unscientific in the mainstream scientific community.

In spite of some scientists referring to the anthropic principle as "the A-word" (said with a smirk) the weak form of the anthropic principle has been used with some success in science. For example, back in 1987, Steven Weinberg placed an upper bound on the size of the cosmological constant using the reasoning that if it were larger, stars and galaxies could not have formed and we would not be here.[17] As mentioned earlier, a lower bound on the cosmological constant also exists from the fact that a large negative cosmological constant would lead the universe to collapse too rapidly for biological and stellar evolution to happen. These bounds amount to a (successful) prediction of the size of the cosmological constant based on anthropic reasoning.

In another example of anthropic science, in 1952 Fred Hoyle inferred the existence of a particular excited state of the carbon nucleus because without it the nuclear fusion processes in the sun would not have been able to produce the amount of carbon that we see in our universe.[18] That excited state of carbon was later found.

In fact, weak anthropic reasoning makes sense in the bubble multiverse—or any other multiverse scenario where there are countless universes sampling all possibilities for the physical parameters that determine the fundamental characteristics of the individual universes. Recall that fluctuations in the eternally inflating reality of the bubble multiverse cause the bubble regions where inflation stops and a universe—or perhaps a beyond-the-horizon-multiverse—grows. The physical characteristics of this bubble region are determined by the details of how the space-time of the extra dimensions is compactified, and there are something like 10^{500} possibilities for how that is done. The universes inhabiting the multiverse sample all different parts of the cosmic landscape. If it *can* happen, it *has* happened. If all possibilities have happened, it's only natural that we'll find ourselves in a universe where we could have evolved to see it *even if that universe is one with a highly improbable set of physical constants.*

Within the multiverse-cosmic landscape scenario, the argument for intelligent design falls flat. We may live in a universe with fine-tuned parameters that allow us to evolve in our present form, but, in the multiverse, all possibilities are to be found and it follows naturally that we find ourselves in a fine-tuned universe. Without that, we wouldn't be there to see it.

Personally, I think the whole God or not-God thing is still a matter of faith. The cosmic landscape-multiverse scenario certainly allows for fine-tuning without intelligent design. Though interesting, is it really true that fine-tuning implies intelligent design in the first place? In the scientific view of things, we evolved to be the way we are because of the characteristics of our universe. Victor Stenger says it nicely: "The fine-tuning argument would tell us that the Sun radiates light so that we can see where we are going. In fact, the human eye evolved to be sensitive to light from the sun. The universe is not fine-tuned for humanity. Humanity is fine-tuned to the universe."[19]

On the other hand, I don't see that the multiverse really threatens the idea of a Creator. Although fine-tuning does not prove the existence of design and/or God, if you choose to believe in God, why not have faith that God works through the multiverse?

At the end of *The Cosmic Landscape*, Leonard Susskind gives a succinct summary of the situation:

> Those who would look to the Anthropic Principle as a sign of a benevolent creator have found no comfort in [the multiverse-string theory landscape picture]. The laws of gravity, quantum mechanics, and a rich Landscape together with the laws of large numbers are all that's needed to explain the friendliness of our patch of the universe.
>
> But on the other hand, neither does anything in this [picture] diminish the likelihood that an intelligent agent created the universe for some purpose. The ultimate existential question, "Why is there Something rather than Nothing?" has no more or less of an answer than before anyone had ever heard of String Theory.[20]

Of course, it may be that a very strict interpretation of bits of scripture would lead some people to believe that our universe must be the only one and this is hard to reconcile with the multiverse. On the other hand, many people are comfortable in their faith without worries about scientific support, and I don't expect that all these ideas floating around about the multiverse will change that very much.

Don't get the idea that there is unanimity among scientists about the multiverse and anthropic reasoning. Some scientists feel the idea of anthropically selecting among the string theory landscape's ensemble of universes is unscientific. Lee Smolin, for example, considers it "unscientific because it lacks a property necessary for any scientific hypothesis—that it be falsifiable."[21]

Irit Maor, Lawrence Krauss, and Glenn Starkman recently published a paper where they argue that the anthropic principle is based on ignorance rather than knowledge.[22] Because of this,

Any probabilistic inferences one draws using [the anthropic principle] should be carefully interpreted. Not only do we have no fundamental theory that might provide an underlying probability distribution for universes with different cosmological constants, we know very little about the range of fundamental parameters that might allow the evolution of intelligent life in a universe.[23]

They maintain that the most we can really say about the cosmological constant using anthropic limits is that "the existence of us and the existence of the observed value of the cosmological constant do not contradict each other."[24]

Lawrence Krauss is an old colleague of mine from young faculty days. He's the sort of strongly opinionated guy that you'd love to hate. I've heard him described as arrogant. He always thinks he's right. But you just can't help but like Lawrence. He's charismatic, very sharp, quick-witted, and good-humored. *And* he is most often right. He's become a public face for science, and I'm glad of that for science and humanity.

Krauss is on record as being quite critical of string theory in general—or at least the amount of effort that has been put into it and the degree to which it garners attention. He says,

Particle physics provides very elegant mechanisms for generating the density perturbations in the very early universe that might ultimately collapse to form galaxies of visible and dark matter. It is not clear that the additional intellectual overhead associated with branes and extra dimensions is needed to explain anything that we might otherwise explain without it.[25]

According to Krauss, the idea of explaining how the universe behaves as it does using the string theory landscape coupled with anthropic reasoning

G oes completely against the grain of the entire history of physics over the past four hundred years. ...if the landscape turns out the be the main physical implication of the grand edifice of string theory or M-theory, then instead of precise predictions about why the observable universe of three large and expanding spatial dimensions must be the way it is, we might be left with the mere suggestion that anything goes. What was touted twenty years ago as a Theory of Everything would then instead have turned quite literally into a Theory of Nothing.[26]

Skating around on the perilously thin ice of the anthropic principle and intelligent design and in and out amongst the string theory wars is getting wearisome. Let's return to the solid footing that served as the starting point of our journey, the extended Copernican revolution, and ponder it in light of the multiverse. By moving the Earth from the center of the universe to a place orbiting the sun in a heliocentric universe, humanity finally understood the movements of the planets and stars in the heavens. The acceptance of change in the living world led to an understanding of the importance of biological evolution and insights into the nature of life. The realization that humans are animals and a product of evolution has given us knowledge about our origins and strengths and weaknesses as a species. Relativity led to the realization that the absolute nature of space and time that we held so dear were not, in fact, a reflection of the true nature of reality. With this realization came a deeper understanding of space-time, the keys to nuclear power and the knowledge of the underlying power source for stars. What seemed to be solid matter is now known to be mostly empty space. In the quantum world, particles are waves and waves are particles, and the deterministic universe seems to be gone forever. Yet, the development of quantum theory has driven many of the amazing scientific and technical advances in the last century. Again and again, science has taken great leaps forward at the expense of human bias.

Now we find out that our *universe* isn't even special. Is the multiverse the next great step in the Copernican revolution? Perhaps. Can we go any further than this away from human bias? Max Tegmark thinks so. He has dreamed up something called the "mathematical universe hypothesis," which is—well, it's Copernicus on steroids.[27]

Tegmark seems to have found the inspiration for the mathematical universe, at least in part, from Eugene Wigner, one of the great theoretical physicists of the last century. In a 1959 essay, Wigner goes through example after example of instances that demonstrate the "unreasonable effectiveness of mathematics in the natural sciences."[28] According to Wigner, "...the enormous usefulness of mathematics in the natural sciences is something bordering on the mysterious and...there is no rational explanation for it."[29]

The concept of Tegmark's mathematical universe hypothesis has a rather innocent foundation. The underlying assumption is that there exists an external reality completely independent from humans. Most scientists would agree with this as it underlies the basic methodology of science. Strict believers in the Copenhagen interpretation of quantum mechanics and believers in the multiverse of wishful thinking would object to it, because they feel that the act of observation and/or the mind plays a critical role in bringing about the reality we experience.

The belief in an external reality is closely tied with the physicists' pursuit of the theory of everything. In fact, there's hardly a point to seeking such a theory of everything if it doesn't describe an independent external reality that we all experience. This theory of everything has been the goal of physics for 400 years, and we owe a great deal of our scientific knowledge to the push in that direction.

For now, let's assume there is an external reality independent of humans and that there is such a thing as a theory of everything to be discovered. What possible form can this theory of everything take? It is, after all, a theory of *everything*. According to Tegmark:

> f we assume that reality exists independently of humans, then for a description to be complete, it must also be well-defined according to non-human entities—aliens or super-computers, say—that lack any understanding of human concepts. Put differently, such a description must be expressible in a form that is devoid of any human baggage like "particle," "observation" or other English words.[30]

Enter mathematics. Most of us think of math and yawn. It's a way to calculate a tip after a dinner out or the size of our grocery bill. My physics students probably think of it as a torturous way to spend an evening doing practice problems. To a mathematician, mathematics looks quite different. It is a set of relations between abstract objects. The mathematics we all grow up hating in grade school wasn't handed to us on a stone tablet. Rather, it is based on a set of concepts, relations, and rules that have been developed and discovered over many centuries. Though what we learn in school is particularly useful, mathematicians have discovered many other mathematical structures. In fact, it's not at all unusual for an area of mathematics to be very well developed and studied by mathematicians long before a practical use is found for that area. One particular example that comes to mind is the area of differential geometry, on which Einstein's general theory of relativity is based. The mathematics of differential geometry existed long before Einstein realized it suited his needs as the foundation of a new physics theory.

You can imagine debating whether mathematics consists of relationships *discovered* by humans or *invented* by humans. I fall in the discovery camp, as does Max Tegmark, who says, "We discover [mathematical structures], and only invent the notation for describing them."[31] The belief that mathematical structures exist independent of whether or not they've been discovered by humans is quite similar to the belief in an external physical reality. Physics and mathematics are quite different, however. Whereas physics is constrained to describe reality, mathematics is constrained only to be internally consistent. In a mathematical structure, the parts and rules are put forth, and everything else follows from logical consistency. There's no constraint that the mathematics be useful or describe our particular reality.

Though mathematics is not burdened with the requirement to describe reality, mankind, at least as far back as the time of Pythagoras in ancient Greece, has long sought to describe reality through mathematics. Reasons for this include the precision, the lack of ambiguity, and the beauty of mathematics, as well as the power of mathematics to help us comprehend and describe things that are lost on us otherwise. Though perhaps the unreasonable effectiveness of mathematics in describing reality is puzzling, it has long been recognized.

Tegmark takes things a bit further with his ideas. In an interview with my Rochester colleague Adam Frank, published in *Discover Magazine*, Tegmark says:

> Galileo and Wigner and lots of other scientists would argue that abstract mathematics "describes" reality. Plato would say that mathematics exists somewhere out there as an ideal reality. I am working in between. I have this sort of crazy-sounding idea that the reason why mathematics is so effective at describing reality is that it is reality. That is the mathematical universe hypothesis: Mathematical things actually exist, and they are actually physical reality.[32]

Yes, Max, it *is* a crazy-sounding idea. Let's look at it a bit more closely. We start from the existence of an external reality independent of humans, and we assume there exists a theory of everything to describe our universe. Tegmark says that such a theory of everything must be a self-contained and consistent mathematical structure. Otherwise, it would contain baggage that would be open to interpretation. Because mathematical structures exist regardless of whether or not they've been discovered by humanity and they consist only of objects and the relations between them, mathematics is the ideal form for the theory of everything.

Now comes what most of us would think of as the strange part of Tegmark's argument. A theory of everything is just that: a theory of *everything* in our universe. That means the theory has to have a unique bit of mathematics that corresponds to every little physical thing that exists in our universe and vice versa. There exists a one-to-one correspondence between aspects of our physical universe and bits of this mathematical theory of everything. A mathematician would call this an "isomorphic mapping" between the universe and the theory of everything. This means the two things are completely equivalent. Our universe *is* a mathematical structure.

Now, if our universe has an existence as physical reality *and* a mathematical structure, what makes it special? Tegmark suggests "that complete mathematical democracy holds: that mathematical existence and physical existence are equivalent, so that all mathematical structures exist physically as well."[33] This—presumably infinite—set of mathematical structures and the physical manifestations of them, constitutes a multiverse. Each independent mathematical structure corresponds to a unique physical universe in the greater reality. This is what Tegmark calls the Level IV multiverse. In the taxonomy I introduced at the start of the book, I have called this multiverse the "mathematical multiverse."

I think of the mathematical universe hypothesis as Copernicus on steroids, because I can't conceive how we can move any further away from the human bias than this. According to Tegmark, "the hypothesis predicts a lot more to reality than we thought, since every mathematical structure is another universe. Just as our sun is not the center of the galaxy but just another star, so too our universe is just another mathematical structure in a cosmos full of mathematical structures."[34] We aren't talking about a universe that simply differs from ours by some varying physical constants. All the rules and the way things are put together differ between the unique structures/universes. Basically, anything can happen so long as there is logical consistency within each universe. Compared to the mathematical multiverse, the string theory landscape seems manageable. In fact, it is this way for a good reason. With Tegmark's mathematical democracy proposal, the mathematical multiverse contains all the others. He says, "...any conceivable parallel universe theory can be described at Level IV."[35] He calls the mathematical multiverse "the ultimate Ensemble theory" since "it subsumes all other ensembles, therefore bring[ing] closure to the hierarchy of multiverses, and there cannot be say a Level V."[36]

Because we've already broached the topic of God and the multiverse in this chapter, how does God fit into the mathematical multiverse? It may be that the extreme inclusiveness of the mathematical multiverse brings about an odd convergence. Paul Davies at Arizona State University argues that Tegmark's mathematical universe hypothesis and deism "will turn out to be of equivalent complexity because *they are contained within each other*."[37] He goes on to clarify this statement, saying:

Consider the most general multiverse theories (e.g. Tegmark's Level IV version [or the mathematical multiverse]), where even laws are abandoned and anything at all can happen. At least some of these universes will feature miraculous events…. They will also contain thoroughly convincing religious experiences, such as direct revelation of a transcendent being. It follows that a general multiverse set must contain a subset that conforms to traditional religious notions of God and design. It could be countered, however, that this subset is embedded in a much bigger set in which no coherent theological plan is discernible, so that a random observer would be unlikely to encounter a world in which a god was seen to be at work."[38]

I categorize the mathematical multiverse as a faith-based multiverse, in part, because it hinges on several beliefs: the existence of an external reality independent of humans, the existence of a theory of everything, and the existence of mathematics outside of human invention. I think all of these beliefs are quite reasonable in the eyes of modern science and mathematics. Max really isn't crazy.

Tegmark suggests that evidence for the mathematical universe can be found possibly in the degree to which our universe is typical.[39] The idea is that if we could calculate the overall probability distribution for some fundamental aspect, such as the cosmological constant or the force of gravity, of "the multiverse where that quantity is defined" and we were able to show that what we observe in our universe is *not* typical, it would rule out the mathematical multiverse. In principle, Tegmark is right about this, and it makes the mathematical multiverse falsifiable.

Tegmark's argument about the falsifiability of the mathematical multiverse is vulnerable to the argument made by Maor, Krauss, and Starkman against anthropic reasoning that I mentioned earlier[40]— namely that in the absence of the real theory we don't know the probability distributions of fundamental physical constants in the multiverse. For example, in the vast multiverse, over what range of values will we find the cosmological constant, or the strength of the force of gravity? In this case, where we are considering the mathematical multiverse, the argument posed by Maor, Krauss, and Starkman is even stronger. To get a handle on any probability distribution for a physical constant, we'd need to know not just *one* theory, but *all the mathematical structures in the multiverse*—at least those where the quantity under study is defined. The number of theories to consider must be significantly larger than the number of possibilities in the string theory landscape, which is subsumed by the mathematical multiverse. In addition, how would we ever *know* that we have the right theory? We could postulate a theory or a group of theories and calculate a probability distribution for a physical constant and possibly show that our universe is atypical for that distribution. But, if our theories are incorrect or incomplete, there's no way we could know. Consequently, I see this multiverse as more faith-based than scientific.

Perhaps now is a good time to reread Chapter 1 and glance back at Figure 1.1 and Figure 1.2. Ponder our journey through the multiverses. Or is it multiversae? There's something perverse about writing a word that represents the plural of multiple universes. Talk about the ultimate plurality!

In this book we have discussed multiverse concepts where the universes within them are separated by space-time: the fecund multiverse, the cyclic patch multiverse, the bubble multiverse, the cyclic big bang multiverse, the ekpyrotic multiverse, and the beyond-the-horizon multiverse; concepts where the universes within them are separated by dimensions: the many-worlds multiverse; and concepts that are either best described as a matter of faith or not so easily categorized in terms of separation by a physical quantity: the computational multiverse, the multiverse of faith, the multiverse of wishful thinking, and the mathematical multiverse. I imagine this list is incomplete, and I look forward to hearing of other candidates for inclusion in this taxonomy.

I hope that I've convinced you that *the multiverse is here to stay*. This is not a passing fad. In fact, many of these ideas are not even new. What is new is the inference of multiverse concepts from the most advanced scientific theories of our day.

Are all these inferred multiverses *really* science? That's a question that's under vigorous debate. It is quite possibly something that will never be settled satisfactorily. Personally, I think the inference of the scientific multiverse concepts is based in sound science, but I'm not holding my breath until we have some form of solid confirmation or falsification. Still, time and time again, we have run up against the immeasurable only to find that 30 years down the line a new technology or a very smart person opens up the previously hidden realm for exploration. I'd like to think that may be the case here. The issue is, of course, that by construction we can't go out and see other universes in the multiverse, but, as people continue to probe these questions, perhaps they'll find a different way to say something.

There are ongoing scientific efforts that may shed light on things that are relevant for the multiverse. For example, at the LHC, physicists are looking for evidence of extra dimensions at the extraordinarily short distances probed by those collisions. Plans are underway to use distant supernovae to study the expansion history of the universe with greater precision and further back in time than has been done previously. This may tell us more about the nature of dark energy. If ongoing searches for dark matter meet with success, that will help pin down one of the uncertainties in our cosmological models.

The degree of speculation implicit in the different scientific multiverse concepts is not equal. The space-separated regions of the expanding universe—the beyond-the-horizon multiverse—is not particularly controversial among cosmologists. When pressed, I think most physicists would agree that it is there. Inflation seems compelling in modern cosmology, and is supported by growing direct and indirect evidence. If inflation really happened, the beyond-the-horizon multiverse is certainly there and the bubble multiverse seems difficult to avoid. Quantum mechanics works quite well, and quantum mechanical processes can be understood through the elegant dimensionally-separated many-worlds multiverse. These are all concepts with strong foundations in mainstream science.

Are all multiverses here to stay? Not at all. Inflationary cosmology and the ekpyrotic multiverse are probably mutually exclusive. Smolin's fecund multiverse will live or die by our eventual understanding of quantum gravity and the degree to which our universe is optimal for black hole production The multitude of religious multiverse concepts are here to stay, but they are destined to remain a matter of faith and not science.

The categorizations in Appendix A are hardly exclusive, or even unique, for most of these multiverse concepts. The mathematical multiverse, if it exists, contains within it all the other multiverses. The ekpyrotic multiverse consists of cycles separated by time—if we can argue that time exists outside the brane collisions somehow—and yet there is a dimensional separation between the branes that is critical to the theory. A brane in the ekpyrotic universe contains a beyond-the-horizon multiverse. Because quantum mechanics is thought to pervade the cosmological multiverses that we've discussed, the many-worlds multiverse is present in each of the cosmological multiverses. Or, perhaps it's better to think of the great wave function in the infinite-dimensional Hilbert space in the many worlds multiverse as including all the bits of the cosmological multiverse(s). Is there quantum mechanics in Heaven? Within the computational multiverse, are there simulations of the quantum mechanical wave-function that lead to a many-worlds multiverse? Surely there are computational multiverses in different regions of the beyond-the-horizon multiverse.

Should you believe in the multiverse? Steven Weinberg tells the story of how he picked up an issue of *Astronomy* magazine in the Austin airport and a story on multiverses caught his eye. Weinberg says, "Inside I found a report of a discussion at a conference at Stanford, at which Martin Rees said that he was sufficiently confident about the multiverse to bet his dog's life on it, while Andrei Linde said he would bet his own life. As for me, I have just enough confidence about the multiverse to bet the lives of both Andrei Linde *and* Martin Rees's dog."[41]

Should you believe in the multiverse? That's for you to decide. I confess that I am much more the believer now than I was before I wrote this book. Before, I was dubious about the whole business. Now, I think the question is not "Do we live in a multiverse?" Rather the questions are "Do we live in multiple types of multiverses?" and, if so, "Which ones?"

Appendix A
An Overview of Multiverse Concepts

Space-Time Separated Multiverses

1 Beyond-the-Horizon Multiverse

The space in the cosmos appears to be flat and vastly larger than our potentially observable universe. This is particularly true in inflationary cosmologies. Distant parts of the cosmos—forever causally disconnected from us by the expanding universe—will have physical laws similar to ours but different initial conditions.

2 Fecund Multiverse

It has been suggested that quantum gravity effects inside black holes might lead to the birth of another universe isolated by the black hole from the universe that spawned the black hole. The new universe is born with physical laws and constants rather similar to the parent universe, leading to the idea of cosmological natural selection where universes more favorable to black hole formation will be selected by a form of multiverse evolution. Interestingly, conditions for black hole formation tend to be favorable for star formation and the existence of life as we know it.

3 Oscillating Big Bang Multiverse

Though it is now thought not to be a viable scientific model, for many years cosmologists entertained the idea of an oscillatory "big bang–big crunch" multiverse. In this scenario, the universe expands after a big bang, then slows and contracts in a big crunch only to repeat the cycle again and again.

4 Ekpyrotic Multiverse

A cyclic universe concept has been spawned from string theory concepts. In this cosmology, two three-dimensional structures called branes are separated by a extra dimension. These branes collide, creating what appears as the initial conditions of the big bang within each brane. As the branes move apart after the collision, the space within each brane expands and the matter cools, creating something similar to what we observe in the universe today. The expansion within each brane continues until the branes, still wafting around in the extra dimension, collide once again, creating an apparent big bang within the branes and starting the next cycle in the multiverse.

5 Cyclic Patch Multiverse

A different oscillatory model has been put forth that contains string theory concepts, as well as elements of both the oscillating big bang multiverse and the Ekpyrotic universe models. In this model, for a universe such as ours, dark energy accelerates the expansion of the universe to such an extent that the matter density effectively drops to zero, avoiding the technical problems with entropy plaguing the oscillating big bang multiverse. At some point, the hyper-expanding universe reverses direction and each—much smaller—causally disconnected region within the original expanded universe, contracts into a big crunch and then bounces to form a new big bang. It is as if the universe expands into a great quilt and fragments, with each patch of the quilt contracting and bouncing into a new universe. This cycle continues indefinitely.

6 Bubble Multiverse

In big bang inflationary cosmology, it is hypothesized that the inflationary hyper-expansion of the cosmos is eternal with the conditions in different domains settling down and creating pocket (or bubble) universes throughout the inflating cosmos. Each pocket universe might have a different dimensionality and a different set of physical constants. In the language of string theory, each pocket universe lies in a different spot in the landscape of possibilities that string theory allows.

Dimensionally Separated Multiverses

7 Many-Worlds Multiverse

In quantum mechanics, different quantum mechanical outcomes are separated in a mathematical space called Hilbert space. In the many worlds interpretation of quantum mechanics coupled with the concept of decoherence, these different realities emerge as distinct branches of the evolving universal wave function—effectively parallel universes.

Faith-Based Multiverses

8 Multiverse of Faith

Humans have always had concepts of other planes of existence, often tied up with religion and concepts of the afterlife. The lack of scientific evidence supporting these planes of existence has had little effect on the adherents. After all, this is a multiverse based on faith.

9 Mathematical Multiverse

A multiverse based on mathematical reality has been proposed. In this case, faith in an external reality and a theory of everything lead to the idea that fully consistent and self-contained mathematical descriptions can correspond to different universes. The individual universes in this multiverse can have completely different physical laws. Though this multiverse concept is placed in the faith-based class, scientific arguments form the framework of the concept, and further work may shed light on its viability.

10 Computational Multiverse

If you believe that we are not the only intelligent life in the universe, it isn't much of a stretch to imagine that there might be beings that are vastly more intelligent and capable than we. Such beings might have learned how to simulate universes—stars, life, consciousness, and all—on some super-powerful alien computer. The simulation might be so good as to allow simulated beings to simulate universes. Simulations could be made using all possible sets of initial conditions and physical constants. Before long this line of thinking leads to an infinite multiverse of simulated universes. Welcome to *The Matrix*.

11 Multiverse of Wishful Thinking

One of the current snake-oils on the market is something called the Law of Attraction, which goes something like "if you want something enough and really believe in it, the universe will deliver it to you." The basic idea is that desire in the mind can control one's flow among the many dimensionally separated universes of quantum mechanics. This idea is loosely based in the many-worlds interpretation of quantum mechanics, but is placed in the faith-based class because the Law of Attraction is not scientifically supported. People may choose to believe in it just as they may choose to believe in a particular version of Heaven.

Appendix B
Tegmark's Taxonomy of the Multiverse

Tegmark's Level I Multiverse

A cosmic reality vastly larger spatially than our observable universe—consistent with the observed uniformity and geometrical flatness of space in our observable universe—has an infinite number of regions forever destined to be beyond the cosmic horizon. This means that these regions will never be part of our observable universe and are causally disconnected from our universe. These regions would have similar physical laws but different initial conditions from our observable universe. A Level I multiverse is the same as the space-time separated beyond-the-horizon multiverse in Appendix A.

Tegmark's Level II Multiverse

In models with eternal cosmic inflation, the process of inflation never stops. Rather, inflation ends in some domains of the inflating reality while continuing elsewhere. Each of the regions where inflation ends forms a bubble (or pocket) universe with potentially differing dimensionality and physical constants. Each bubble might become a Level I multiverse in its own right. A Level II multiverse corresponds to the spatially separated bubble multiverse in Appendix A.

Tegmark's Level III Multiverse

In quantum mechanics, reality is a universal wave function in Hilbert space containing a number of possible emergent realities branching out into the larger whole. In the many-worlds interpretation of quantum mechanics, the reality we perceive is just one potential path within the countless possible paths within the universal wave function. These different emergent paths can be thought of as parallel universes. A Level III multiverse is the same as the dimensionally separated many-worlds multiverse in Appendix A.

Tegmark's Level IV Multiverse

Many scientists have faith in external reality and the existence of a theory of everything. Assuming the external reality and the theory of everything, Tegmark argues, leads to a mathematical reality. According to this mathematical universe hypothesis, all possible self-contained and self-consistent mathematical structures inherently describe a universe in a sort of a one-to-one mapping. Theses universes can have very different physical laws as well as different physical constants, particles, and so forth. The Level IV multiverse corresponds to the faith-based mathematical multiverse in Appendix A.

Notes

Introduction

1. Oops. No offense intended to you Trekkies. Some of my best friends are Trekkies. Live long and prosper and all that.

2. See the "Parallels" episode of *Star Trek: The Next Generation*, for example.

Chapter 1

1. Epicycles are circles on circles. In order to account for the observed motion of the planets it was necessary to modify the nested sphere picture by having planets move on small circles whose centers moved around the Earth on a larger sphere rotating about the Earth.

2. Singham (2007).

3. This last statement is a bit contentious. After all, we evolved and survived in this universe, and that required a particular set of physical laws and physical constants. We'll discuss this issue more in the latter part of the book.

4. Professor Tegmark has discussed his multiverse taxonomy in a number of papers. For a nice discussion, see Tegmark (2004).

5. Just kidding, of course. MIT is a great university that can boast of having many talented, brilliant faculty members. Why, some of my dearest friends are MIT faculty or alums.

6. Frank (2008).

Chapter 2

1. Supersymmetry is the name of a fundamental symmetry between particles called fermions and particles called bosons. It's exhibited in many theories of the particles and forces of nature. There are compelling reasons to think it might be true though there is no direct experimental evidence for its existence. More on this in subsequent chapters. The details of this are not important for the current story.

2. There still isn't any direct evidence for supersymmetry, though the theoretical case for it is currently even more compelling than it was during my days in graduate school. Many of us physics types (including me) are hopeful that the first direct evidence of supersymmetry will be discovered at the Large Hadron Collider in the next few years.

3. Susskind (2006), p.112.

4. Ibid., p.114.

5. Johnson (2008).

6. Susskind (2006), p.113.

7. Okay. Perhaps not *all* of us. I understand the Australian Aborigines have a rather different concept of time and I've certainly wondered on occasion about my significant other's concept of time—though I think elaborating on that here would be unwise.

8. Clark (1971), pp. 113–118. In spite of what I say here, it is not clear that the Michelson and Morley results played a significant role in Einstein's thinking. He was bothered by other fundamental issues concerning light. For example, it is known that he wondered how a ray of light might appear to an observer moving at the speed of light.

9. Albert Michelson won the 1907 Nobel Prize in Physics, the first U.S. citizen to do so.

10. These transformations are called the Lorentz transformations after a physicist named Hendrik Lorentz. Though Einstein is given credit appropriately for inventing the ideas of special relativity, he didn't pull it all out of thin air. Several scientists prior to Einstein, such as Lorentz, George FizGerald, and Henri Poincaré, had already discussed some aspects of what became special relativity. The interested reader is referred to Kaku (2004), Rigden (2005) and Pais (1982).

11. String theory is a geometric theory where everything is constructed of different types of structures in space-time. More on this in Chapter 8.

12. Momentum can be defined as mass multiplied by velocity in physics. Intuitively, it can be thought of exactly in the way the word is used in common language. In physics, however, it has a formal definition as mass multiplied by velocity and has both a magnitude and direction. Momentum is conserved in an isolated system, which is to say that the sum of the incoming momenta is equal to the sum of the outgoing momenta. Momentum conservation is a powerful analytical tool in physics.

13. Clark (1971), p. 86.

14. This paper also dealt with electromagnetism, or light. In fact, the paper was entitled "Zur Elektrodynamik bewegter Körper," which is translated as "On the Electrodynamics of Moving Bodies" in English.

15. A "g" represents an acceleration of 9.8 meters/(second)^2, which is the rate at which dropped objects accelerate downward at the surface of the Earth.

16. Will (2006).

17. This point might be a bit contentious over drinks with physicists because there is a form of radiation emitted from the edge of black holes called Hawking radiation.

Chapter 3

1. Goldman (1983), p. 152.
2. Goldman (1983).
3. Ibid., p.34.
4. Ibid., p. 61.
5. Lewis (1926).
6. O'Connor and Robertson (Planck, 2003).
7. O'Connor and Robertson (de Broglie, 2001).
8. Schrödinger was awarded the 1933 Nobel Prize in Physics for this work.
9. Heisenberg received the 1932 Nobel Prize in Physics for his invention of quantum mechanics.
10. According to Pais, Heisenberg made the initial breakthrough and Born and Jordan contributed by cleaning up the mathematics. See Pais (1991), pp. 275–279.
11. Pais (1991), p. 281.

Chapter 4

1. Pais (1991), p. 306.
2. Clark (1971), p. 414.
3. Proponents of the Copenhagen school of thought on quantum mechanics have not been unanimous in their opinions on important questions. For example, there is disagreement among physicists as to the degree to which the wave function is something that is physically real, as opposed to a tool with which to understand probabilities in a quantum mechanical problem.
4. For more strange quantum fun, the interested reader is urged to look up the work of scientists David Bohm, Alain Aspect, John Bell, Len Mandel, and John Clauser.
5. Zurek (2003).

6. Tegmark and Wheeler (2001).

7. Byrne (2007).

8. Ibid., p. 102.

9. Zurek (2003) and Tegmark and Wheeler (2001).

10. See, for example, discussions in Zurek (2003) and Tegmark and Wheeler (2001).

11. Zurek (2003).

12. Tegmark (2004).

13. Wallace (2010).

14. Albert and Loewer (1988), and Lockwood (1996).

15. Albert and Loewer (1988), p. 203.

16. See, for example, Penrose (1994), Squires (1990), and Herbert (1993).

17. Stenger (1995).

18. F.A. Wolf, in *What the Bleep Do we Know!?* (2004).

19. A.Goswami, in *What the Bleep Do we Know!?* (2004).

20. *What the Bleep Do we Know!?* (2004).

21. Byrne (2006).

22. Ibid., p. 4.

23. Ibid., p. 15.

24. Ibid., p. 21.

25. Ibid., p. 160.

26. Goswami (2010).

27. Ibid.

28. Tegmark (2000).

29. See, for example, Asitin (2003).

Chapter 5

1. "The Nobel Prize in Physics 1965," Nobelprize.org, August 17, 2010, *nobelprize.org/nobel_prizes/physics/laureates/1965/*.

2. Richard Feynman, in a speech to the 1966 National Science Teachers' Meeting as quoted in Feynman (1999), p. 187.

3. See Feynman (1986) for a video clip of Professor Feynman talking about the O-ring during the televised hearings of the Rogers Commission in 1986.

4. See Quinn and Witherell (1998) and Aitchison, Cowan, and Long (2008).

5. For more information about particle physics, see the following: *The Particle Adventure*, Particle Data Group of Lawrence Berkeley National Laboratory; Riordan (1987); and Lederman and Teresi (1993).

6. This onion analogy is commonly used among particle physicists. Frank Close has written a nice book on particle physics entitled *The Cosmic Onion*. If you are looking for something that describes the nature of matter in a fashion that is somewhat more technical that you'll find here and in my previous suggestions, you might look at the update of Close's classic, Close (2007).

7. 0.00002 meters is 20 millionths of a meter and is typically called 20 micrometers, or 20 μm.

8. 10^{-10} meters if you prefer scientific notation.

9. The LHC is expected to probe down to roughly 10^{-19} meters eventually.

10. Cosmic rays continue to be studied with enthusiasm. Rare cosmic rays have energies far greater than physicists can achieve in manmade accelerators. There is interest in knowing the sources of these particles as well as the acceleration mechanism. Recently, physicists using the Pierre Auger Observatory in western Argentina collected data with enough

directional information to infer that large black holes in the nearby active galaxies are the likely sources of the highest energy cosmic rays. See Pierre Auger Collaboration (2008).

11. Riordan (1987), p. 49.

12. Baeyer (1993).

13. "The Nobel Prize in Physics 1944," Nobelprize.org, August 18, 2010, *nobelprize.org/nobel_prizes/physics/laureates/1944/*.

14. A nice history of particle physics from this era through the 1970s is given in Riordan (1987).

15. Riordan (1987), p. 102.

16. Ibid., p. 210.

17. Kendall, Friedman, and Taylor were awarded the 1990 Nobel Prize in Physics "for their pioneering investigations concerning deep inelastic scattering of electrons on protons and bound neutrons, which have been of essential importance for the development of the quark model in particle physics." See "The Nobel Prize in Physics 1990," Nobelprize.org, August 21, 2010, *nobelprize.org/nobel_prizes/physics/laureates/1990/*.

18. Riordan (1987), p. 176.

19. Burton Richter and Samuel C.C. Ting shared the 1976 Nobel Prize in Physics for the discovery of this particle, called the J-psi. See "The Nobel Prize in Physics 1976," Nobelprize. org, August 21, 2010, *nobelprize.org/nobel_prizes/physics/laureates/1976/*.

20. Riordan (1987), p. 282.

21. Weak neutral currents are weak nuclear interactions involving the exchange of the electrically neutral Z particle. For more information, see Haidt (2004).

22. The R-crisis was the name given to the problem seen in the early 1970s where the total number of events produced in electron-positron collisions did not agree with expectations from any of the theoretical models. Theoretical agreement

was achieved finally with the development of quantum chromodynamics as a theory of the strong nuclear interaction. See Pickering (1984), pp. 254–261, and Riordan (1987), pp. 258–261.

23. A. Pickering, *Constructing Quarks*, The University of Chicago Press, 1984, p. 254.

24. "The Nobel Prize in Physics 2004," Nobelprize.org, August 21, 2010, *nobelprize.org/nobel_prizes/physics/laureates/2004/.*

25. Weinberg, Salam, and Glashow shared the 1979 Nobel Prize in Physics and Veltman and t'Hooft shared the 1999 Nobel Prize in Physics. See "The Nobel Prize in Physics 1979," Nobelprize.org, August 21, 2010, *nobelprize.org/nobel_prizes/physics/laureates/1979/* and "The Nobel Prize in Physics 1999," Nobelprize.org, August 21, 2010, *nobelprize.org/nobel_prizes/physics/laureates/1999/.*

26. It's not enough to simply add the new particle. It is also necessary to consider the situation of the Higgs being in a particular ground state even though there are many other, equally probable possibilities. This theoretical procedure is known as electroweak symmetry breaking.

27. Carlo Rubbia and Simon van der Meer won the 1984 Nobel Prize in Physics for this discovery. See "The Nobel Prize in Physics 1984," Nobelprize.org, August 22, 2010, *nobelprize.org/nobel_prizes/physics/laureates/1984/.* Also, see Taubes (1987).

28. These groups were the ALEPH, DELPHI, OPAL, and L3 collaborations at the LEP collider at CERN and the MARK II and SLD collaborations working at the Stanford Linear Collider at SLAC.

29. I have not yet discussed the discovery of the tau lepton and the bottom and top quarks. These discoveries, though important, were not essential for the development of the early Standard Model paradigm.

30. Martin Perl won the Nobel Prize in Physics for this discovery. See "The Nobel Prize in Physics 1995," Nobelprize.org, August 22, 2010, *nobelprize.org/nobel_prizes/physics/laureates/1995/*.

31. Knill (1998).

32. Soter and Tyson (2001), p. 135.

33. O. Knill, "Supernovae, an alpine climb and space travel," on-line essay, July 14, 1998, *www.dynamical-systems.org/ zwicky/Zwicky-e.html*.

34. Though it is a bit out of date now, see Krauss (1989).

35. I may have overstated the degree to which supersymmetry solves the hierarchy problem. It would likely reduce the Standard Model hierarchy problem and possibly create new questions as to what determines and stabilizes the masses of the supersymmetric particles.

Chapter 6

1. Alphaville Vintage Toys, *www.alphaville.com/*.

2. Clark (1971), p. 270.

3. Spring 2010.

4. Steven Weinberg, Nobel laureate of electroweak fame, wrote a wonderful book that takes the reader on a tour of the early period of the hot big bang model. See Weinberg (1977).

5. See Frank (2007) for a delightful article on this topic.

6. Guth (1997).

7. Bennet, Turner, and White (1997). The authors are well-known cosmologists and this paper is an extremely nice, albeit somewhat technical (aimed at non-expert physicists), introduction to the issues surrounding the detection and analysis of the cosmic microwave background radiation.

8. Happer, Peebles, and Wilkinson.

9. After correcting for the Doppler effect from Earth's absolute motion through the cosmos.

10. Wilson (1978).

11. Ibid.

12. "Professor George Smoot," cited on November 2, 2010), *aether.lbl.gov/www/personnel/Smoot-bio.html.*

13. "The Nobel Prize in Physics 2006," Nobelprize.org, November 2, 2010, *nobelprize.org/nobel_prizes/physics/laureates/2006/.*

14. "Wilkinson Microwave Anistropy Probe," NASA.gov, November 2, 2010, *map.gsfc.nasa.gov/.*

15. See the discussion in Steinhardt and Turok (2007), pp. 180–182.

16. Ostriker and Steinhardt (2001).

17. See Baum and Frampton (2007) for the technical report. For the rest of us, see Atkins (2007) for a nice description of the theory.

18. Each patch undergoes inflation at the bounce, which is a point I omit here because I won't cover inflation until Chapter 7.

19. Much of this evidence is well discussed in Tegmark (2002).

20. Actually, physicists call this the particle horizon, but I'll ignore this for now, because I'd prefer not to introduce too many terms.

21. Lawrence Krauss as quoted in Klotz (2009).

22. Krauss and Starkman (1999).

23. This picture does not yet include the possibility of inflation.

24. Tegmark (2004).

Chapter 7

1. Guth (1997), p. 2.

2. I'm omitting the magnetic monopole problem, which played an important role historically in the development of the concept of inflation. The story of the magnetic monopole

problem and how it helped prompt the invention of inflationary cosmology is related very nicely in Guth (1997).

3. I'm taking the rough numbers for the inflationary evolution of the universe from Guth (1997). I've seen variation in the numbers in works by other reputable cosmologists, and most of them warn that the numbers are not well determined.

4. Greene (2004).

5. Guth (1997).

6. Vilenkin (2006), p. 63.

7. The workshop was the 1982 Nuffield Workshop on the Very Early Universe.

8. Guth (1997), pp. 172–173, 289.

9. This is explained very nicely in Appendix A of Guth (1997), p. 289.

10. Guth (2002).

11. Garriga and Vilenkin (2001).

12. Ibid.

13. Tegmark (2004).

14. Ibid.

15. Garriga and Vilenkin (2001).

16. Guth (1997), p. 245, and Vilenkin (2006), p. 80.

17. See Bousso and Polchinski (2004) and Tegmark (2004).

18. Stenger (2004).

19. Tegmark (2004), and Garriga and Vilenkin (2001).

20. Tegmark (2004).

21. Smolin (2007).

22. Smolin (2006).

23. Tegmark argues that multiverse models grounded in solid modern physics are testable and falsifiable in his review of the subject in Tegmark (2004).

24. Garriga and Vilenkin (2001).

25. Of course, someone may come along and invent a cosmological theory that does not include inflation that can produce a power spectrum of CMB fluctuations similar to those from inflationary models.

26. See the WMAP project Website at *map.gsfc.nasa.gov/*.

27. As related in Steinhardt and Turok (2007).

Chapter 8

1. Busza, Jaffe, Sandweiss and Wilczek (1999).

2. Ellis, Giudice, Mangano, Tkachev and Wiedemann (2008).

3. A great deal is known about the end stages of the stellar life cycle. However, this fascinating topic is not so important to this book. So, I am glossing over a great deal of interesting detail here. For more information on this subject see *aspire. cosmic-ray.org/labs/star_life/starlife_main.html* and *map. gsfc.nasa.gov/universe/rel_stars.html*.

4. Smolin (2007), p. 335.

5. Ibid.

6. Smolin (1997).

7. Smolin (1997), p. 93.

8. Stenger (2004).

9. Smolin (1997), p. 96.

10. Smolin (2007).

11. Silk (1997).

12. Wheeler (2006).

13. Greene (1999).

14. Smolin has written a book that is critical of the degree of attention given to string theory by the physics community. It has fostered interesting discussions among physicists. See Smolin (2006).

15. Susskind (2006), p. 207.

16. Nambu has won just about every award a theoretical physicist can win, including the 2008 Nobel Prize in Physics.

17. Greene (1999), p. 138, and Randall (2005), p. 287.

18. Susskind (2006), p. 229.

19. Really, what is important here is the mass per unit length of the string, but that's a detail.

20. Greene (1999), p. 143.

21. Ibid., p. 163.

22. Randall (2005), p. 288.

23. Randall (2005), p. 262, and Greene (1999), p. 181.

24. Schwarz and Green discovered that string theory in 10 dimensions could be formulated to be *anomaly free*. See Randall (2005), pp. 289–291 for more information on this.

25. Isaksson (1985).

26. Randall (2005), p. 34.

27. Greene (1999), p. 188, and Susskind (2006), p. 234.

28. Randall (2005), p. 292.

29. Susskind (2006), pp. 276–285.

30. Bousso and Polchinski (2004), p. 81.

31. My apologies to the many talented physicists who have contributed to the development of string theory whose contributions were omitted here. This is not meant to be a systematic history of the subject.

32. Bousso and Polchinski (2004), p. 80.

33. Ibid., p. 81.

34. Ibid., p. 82.

35. Ibid.

36. Susskind (2006).

37. Bousso and Polchinski (2006), p. 79.

38. Tegmark (2004).

39. Richter (2006).

40. Steinhardt and Turok (2007), p. 14 and p. 138.

41. Randall (2005), p. 58.

42. Ibid.

43. Krauss (2005), p. 226.

44. Randall (2005), p. 60.

45. Steinhardt and Turok (2007), p. 15.

46. Ibid., pp. 156–161.

47. Steinhardt and Turok (2007), p. 49.

48. Steinhardt and Turok (2007), p. 49.

49. Lemonick (2004), p. 38.

50. Ibid., p. 41.

51. Steinhardt and Turok, p. 149.

Chapter 9

1. The Sims, created by Will Wright, developed by Maxis, and published by Electronic Arts, 2000. See *thesims.ea.com*.

2. Second Life, developed by Linden Lab, 2003. See *http://secondlife.com/*.

3. Weatherspoon (2003), 425.

4. *The Matrix* (1999).

5. Bostrom (2003), p. 243.

6. Barrow (2007), p. 481.

7. Ibid.

8. Davies (2007), p. 496.

9. Behe (1996) and Dembski (2002).

10. Stenger (2004).

11. Barrow and Tipler (1986).

12. Tegmark (2004).

13. Weinberg (2007).

14. Stoeger (2007).

15. Barrow and Tipler (1986).

16. Stoeger (2007).

17. Weinberg (1987).

18. Stenger (2004).

19. Ibid.

20. Susskind (2006), p. 380.

21. Smolin (2007), p. 323.

22. Maor, Kraus, Starkman (2007).

23. Ibid.

24. Ibid.

25. Krauss (2005), p. 228.

26. Ibid, pp. 236–240.

27. Tegmark (2007), Tegmark (2004), and Tegmark (2008).

28. Wigner (1967).

29. Ibid.

30. Tegmark (2007).

31. Ibid.

32. Frank (2008).

33. Tegmark (2004).

34. Frank (2008).

35. Tegmark (2004).

36. Ibid.

37. Davies (2007).

38. Ibid.

39. Tegmark (2007).

40. Maor, Kraus, Starkman (2007).

41. Weinberg (2007).

Bibliography

Aitchison, I., R. Cowan, O. Long, for the BABAR Collaboration. "B-factories confirm matter-antimatter asymmetry; leads to 2008 Nobel Prize in Physics." December 8, 2008. Accessed August 29, 2010. *www-public.slac.stanford.edu/babar/Nobel2008.htm.*

Albert, D., and B. Loewer, (1988). "Interpreting the Many Worlds Interpretation." *Synthese* 77 (November 1988): 195–213.

Asitin, J., S. Shapiro, D. Eisenberg, and K. Forys, (2003). "Mind-Body Medicine: State of the Science, Implications for Practice." *Journal of the American Board of Family Medicine* 16 (2003): 131.

Atkins, W. (2007). "Big Bang theory gets more competition from an endless cycle universe theory." January 31, 2007. Accessed November 2, 2010. *www.itwire.com/science-news/space/9098-big-bang-theory-gets-more-competition-from-an-endless-cycle-universe-theory.*

Baeyer, H. von (1993). "The Cinderella Particle." *Discover* (December 1993).

Barrow, J. (2007). "Living in a simulated universe." *Universe or Multiverse?* Ed. Bernard Carr. Cambridge, UK: Cambridge Press, 2007.

Barrow, J. and Tipler, F. (1986). *The Anthropic Cosmological Principle.* Oxford, UK: Oxford University Press, 1986.

Baum, L. and Frampton, P. (2007). "Turnaround in Cyclic Cosmology." *Physical Review Letters* 98:071301, 2007.

Behe, M. (1996). *Darwin's Black Box*. New York: Touchstone, 1996.

Bennet, C., M. Turner, and M. White (1997). "The Cosmic Rosetta Stone." *Physics Today* (November 1997).

Bostrom, N. (2003). "Are we living in a computer simulation." *The Philosophical Quarterly*, April, 2003.

Bousso, R., and J. Polchinski (2004)."The String Theory Landscape." *Scientific American*, September 2004.

Busza, W., R.L. Jaffe, J. Sandweiss, and F. Wilczek (1999). "Review of Speculative 'Disaster Scenarios' at RHIC." *www.bnl.gov/ rhic/docs/rhicreport.pdf.*

Byrne, Peter (2007). "The Many Worlds of Hugh." *Scientific American*, December 2007: 98–105.

Byrne, Rhonda (2006). *The Secret*. New York: Atria Books, 2006.

Clark, Ronald (1971). *Einstein: The Life and Times*. New York: Avon Books, 1971. pp. 113–118.

Close, Frank (2007). *The New Cosmic Onion*. Boca Raton, Fl.: Taylor and Francis Group, 2007.

Davies, Paul (2007). "Universes galore: where will it all end?" *Universe or Multiverse?* Ed. Bernard Carr. Cambridge, UK: Cambridge Press, 2007.

Dembski, W. (2002). *No Free Lunch*. Lanham, Md.: Rowman & Littlefield, 2002.

Ellis, J., G. Giudice, M. Mangano, I. Tkachev, and U. Wiedemann (2008). "Review of the Safety of LHC Collisions." *Journal of Physics G: Nuclear and Particle Physics*, 35 (2008) 115004.

Feynman, Richard (1986). "Feynman o-ring junta challenger." *www.youtube.com*. August 17, 2010. *www.youtube.com/ watch?v=KYCgotDV1oc.*

Feynman, Richard (1999). *The Pleasure of Finding Things Out*. Ed. Jeffrey Robbins. Cambridge, Mass.: Perseus Books, 1999.

Frank, Adam (2007). "How the Big Bang forged the first elements." *Astronomy Magazine*, October 2007.

Frank, Adam (2008). "Is the Universe Actually made of Math?" *Discover Magazine*, July 2008.

Garriga, J., and A. Vilenkin (2001). "Many Worlds in One," *Physical Review* D64, 2001, 043511.

Goldman, Martin (1983). *The Demon in the Aether*. Edinburgh: Paul Harris Publishing, 1983.

Goswami, Amit (2010). "Amit Goswami, Ph.D." August 15, 2010. *www.amitgoswami.org/*.

Greene, Brian (1999). *The Elegant Universe*. New York: Vintage Books, 1999.

Greene, Brian (2004). *The Fabric of the Cosmos*. New York: Vintage Books, 2004.

Guth, Alan (1997). *The Inflationary Universe*. Reading, Mass.: Perseus Books, 1997.

Guth, Alan (2002). "Inflation and the New Era of High-Precision Cosmology." *MIT Physics Annual*, 2002.

Haidt, D. (2004). "The discovery of weak neutral currents." *CERN Courier*, October 2004.

Happer, W., P.J.E. Peebles, and D.T. Wilkinson. "Robert Dicke—a biographical memoir." On-line essay. November 2, 2010. *www.princeton.edu/physics/about-us/history/memorable-members/robert-dicke/*.

Herbert, N. (1993). *Elemental Mind*. New York: Dutton, 1993.

Isaksson, Eva (1985). "Gunnar Nordstrom (1881–1923) on Gravitation and Relativity." On-line essay. *www.helsinki.fi/~eisaksso/nordstrom/nordstrom.html*.

Johnson Jr., J. (2008). "Black Holes and Scientific Standoffs." *Los Angeles Times* July 26, 2008.

Kaku, Michio (2004). *Einstein's Cosmos*. New York: W.W. Norton, 2004.

Klotz, I. (2009). "Dark energy to erase Big Bang's fading signal." *DiscoverNews*, February 23, 2009. Accessed November 2, 2010. *www.msnbc.msn.com/id/29353276/.*

Knill, O. (1998). "Supernovae, an alpine climb and space travel." On-line essay, July 14, 1998. Accessed August 29, 2010. *www. dynamical-systems.org/zwicky/Zwicky-e.html.*

Krauss, Lawrence (1989). *The Fifth Essence*. New York: Basic Books, 1989.

Krauss, Lawrence (2005). *Hiding in the Mirror*. New York: Viking, 2005.

Krauss, L., and G. Starkman (1999). "The fate of the universe." *Scientific American*, November 1999.

Lederman, L., and D. Teresi (1993). *The God Particle*. New York: Mariner Books, 1993.

Lemonick, M. (2004). "Before the Big Bang." *Discover Magazine*, February 2004.

Lewis, Gilbert N. (1926). "The Conservation of Photons." *Nature* 118 (1926): 874–75.

Lockwood, M. (1996). "'Many Minds' Interpretations of Quantum Mechanics." *The British Journal for the Philosophy of Science.* 47 (1996): 58–188.

Maor, I., L. Krauss, and G. Starkman (2007). "Anthropics and Myopics: Conditional Probabilities and the Cosmological Constant." *Physical Review Letters* 100(4):041301, 2008.

The Matrix. Dir. Larry Wachowski and Andy Wachowski. Warner Brothers Pictures, 1999.

"The Nobel Prize in Physics." Nobelprize.org: The Official Web Site of the Nobel Prize. *nobelprize.org/nobel_prizes/physics/.*

O'Connor, J.J., and E.F. Robertson (de Broglie, 2001). "Louis Victor Pierre Raymond duc de Broglie." *The MacTutor History of Mathematics Archive.* 2001. Accessed August 29, 2010. *www-history.mcs.st-andrews.ac.uk/history/Biographies/Broglie.html.*

O'Connor, J.J., and E.F. Robertson (Planck, 2003). "Max Karl Ernst Ludwig Planck." *The MacTutor History of Mathematics Archive.* 2003.Accessed August 29, 2010. *www-history.mcs.st-andrews.ac.uk/history/Biographies/Planck.html.*

Ostriker, J., and P. Steinhardt (2001). "The Quintessential Universe." *Scientific American,* January 2001: 46–53.

Pais, Abraham (1982). *Subtle is the Lord, the Science and Times of Albert Einstein.* Oxford, UK: Oxford University Press, 1982.

Pais, Abraham (1991). *Niels Bohr's Times, In Physics, Philosophy and Polity.* Oxford, UK: Clarendon Press, 1991.

Particle Data Group of Lawrence Berkeley National Laboratory. "The Particle Adventure." *www.particleadventure.org/.*

Penrose, Roger (1994). *Shadows of the Mind.* Oxford: Oxford University Press, 1994.

Pickering, Andrew (1984). *Constructing Quarks: A Sociological History of Particle Physics.* Chicago: University of Chicago Press, 1984.

Pierre Auger Collaboration (2008). "Correlation of the highest-energy cosmic rays with the position of nearby active galactic nuclei." *Astroparticle Physics.* 29 (2008): 243.

Quinn, H., and M. Witherell (1998). "The Asymmetry between Matter and Antimatter." *Scientific American,* October 1998.

Randall, Lisa (2005). *Warped Passages.* New York: Harper Perennial, 2005.

Richter, Burton (2006). Letter to the Editor. *New York Times,* January 29, 2006. *www.nytimes.com/2006/01/29/books/ review/29mail.html?_r=3&pagewanted=2.*

Rigden, John (2005). *Einstein 1905.* Cambridge: Harvard University Press, 2005.

Riordan, Michael (1987). *The Hunting of the Quark.* New York: Simon and Schuster, 1987.

Silk, J. (1997). "Holistic Cosmology." *Science,* Vol. 277 (1997): 644.

Singham, Mano (2007). "The Copernican Myths." *Physics Today,* December 2007: 48–52.

Smolin, Lee (1997). *The Life of the Cosmos.* Oxford, UK: Oxford University Press, 1997.

Smolin, Lee (2006). *The Trouble with Physics.* New York: Houghton Mifflin Company, 2006.

Smolin, Lee (2007). "Scientific Alternatives to the Anthropic Principle." *Universe or Multiverse?* Ed. Bernard Carr. Cambridge, UK: Cambridge Press, 2007.

Soter, S., and N. deGrasse Tyson (2001). Ed. "Fritz Zwicky's Extraordinary Vision." *Cosmic Horizons.* New York: The New Press and the American Museum of Natural History, 2001.

Squires, Euan (1990). *Conscious Mind in the Physical World.* Bristol, England: Adam Hilger, 1990.

Steinhardt, P., and N. Turok (2007). *Endless Universe, Beyond the Big Bang.* New York: Doubleday, 2007.

Stenger, Victor (1995). *The Unconscious Quantum.* Amherst, N.Y.: Prometheus Books, 1995.

Stenger, Victor (2004). "Is the Universe fine-tuned for us?" *Why Intelligent Design Fails: A Scientific Critique of the New Creationism.* Eds. M. Young and T. Edis. New Brunswick, N.J.: Rutgers University Press. pp. 172–84.

Stoeger, W. (2007). "Are anthropic arguments, involving multi-verses and beyond, legitimate?" *Universe or Multiverse?* Ed. Bernard Carr. Cambridge, UK: Cambridge Press, 2007.

Susskind, Leonard (2006). *The Cosmic Landscape*. New York: Back Bay Books/Little, Brown and Company, 2006.

Taubes, Gary (1987). *Nobel Dreams*. Random House, 1987.

Tegmark, Max (2000). "The importance of Quantum Decoherence in Brain Processes." *Physical Review* E61 (2000): 4194.

Tegmark, Max (2002). "Measuring Spacetime: From the Big Bang to Black Holes." *Science*, May 2002: 1427.

Tegmark, Max (2004). "Parallel Universes." *Science and Ultimate Reality: quantum theory, cosmology and complexity*. Eds. J.D. Barrow, P.C.W. Davies, and C.L. Harper. Cambridge, UK: Cambridge University Press, 2004. pp. 459–91.

Tegmark, Max (2007). "Shut up and calculate." On-line archive. *arxiv.org*. arXiv:0709.4024v1 [physics.pop.ph]. 2007.

Tegmark, Max (2008). "The Mathematical Universe." *Foundations of Physics* 38:101–50. 2008.

Tegmark, Max, and John A. Wheeler (2001). "100 Years of the Quantum." *Scientific American*, February 2001: 68–75.

Vilenkin, Alex (2006). *Many Worlds in One*. New York: Hill and Wang, 2006.

Wallace, David (2010). "Decoherence and Ontology, or: How I learned to stop worrying and love FAPP." *Many Worlds? Everett, Quantum Theory, and Reality*. Eds. S. Saunders, J Barrett, A. Kent, and D. Wallace. Oxford, UK: Oxford University Press, 2010.

Weatherspoon, Brian (2003). "Are you a Sim?" *The Philosophical Quarterly*, July 2003.

Weinberg, Steven (1977). *The First Three Minutes—A Modern View of the Origin of the Universe.* New York: Basic Books, 1977.

Weinberg, Steven (1987). "Anthropic Bound on the Cosmological Constant." *Physical Review Letters* 59, 2607. 1987.

Weinberg, Steven (2007). "Living in the multiverse." *Universe or Multiverse?* Ed. Bernard Carr. Cambridge, UK: Cambridge Press, 2007.

What the Bleep Do We Know!? Dir. W. Arntz, B. Chasse, and M. Vincente. Perf. Fred Alan Wolf, Amit Goswami, and John Hagelin. Captured Light and Lord of the Wind Films, 2004.

Wheeler, J. (2006). "Quantum ideas. Quantum foam, Max Planck and Karl Popper." Video. *www.webofstories.com/play/9542.*

Wigner, E. (1967). "The Unreasonable Effectiveness of Mathematics." *Symmetries and Reflections.* Eds. Walter Moore and Michael Scriven. Bloomington, Ind.: Indiana University Press, 1967.

"Wilkinson Microwave Anistropy Probe." *NASA.gov.* Accessed November 2, 2010). *map.gsfc.nasa.gov/.*

Will, Clifford M. (2006). "The Confrontation between General Relativity and Experiment." *Living Reviews in Relativity* 9. 2006. Accessed August 9, 2010. *www.livingreviews.org/ lrr-2006-3.*

Wilson, Robert (1978). "The Cosmic Microwave Background Radiation." *Nobelprize.org: The Official Web Site of the Nobel Prize. nobelprize.org/nobel_prizes/physics/laureates/1978/ wilson-lecture.pdf.*

Zurek, Wojciech (2003). "Decoherence, Einselection, and the Quantum Origins of the Classical." *Reviews of Modern Physics* 75 (2003): 715–55.

Index

About the Author

Steven Manly grew up as a free-range college brat in North Carolina. He received an undergraduate degree from Pfeiffer College and PhD in physics from Columbia University in New York City. After moving up the faculty ranks at Yale University, he moved to the University of Rochester, where he now resides and terrorizes students in the introductory physics course sequences. Professor Manly works at high energy accelerators around the world, where his research probes the structure of matter and the forces of nature. He frequently lectures on his research at international scientific conferences and has published more than 150 articles in scientific journals.

Professor Manly's abilities and efforts in teaching and outreach have been recognized nationally and locally. He was named the New York State Professor of the Year in 2003 by the Carnegie Foundation for the Advancement of Teaching, and was honored as the Mercer Brugler Distinguished Teaching Professor at the University of Rochester from 2002 to 2005. In 2007, he was named the recipient of the Excellence in Undergraduate Teaching Award by the American Association of Physics Teachers.

Recently, Professor Manly completed *Relativity and Quantum Physics for Beginners* (Steerforth Press, November 2009), which is a light-hearted, graphical look at two of the most exciting intellectual developments of the 20th century. In *Visions of the Multiverse*, Professor Manly takes the reader on a tour of different multiple universe reality concepts and explains in accessible terms why many leading physicists are convinced that we are in the midst of a new Copernican revolution.